An Approach to Improvement that Works

With an Emphasis on the White-Collar Area

by

A. Donald Stratton
Manager, Quality, AT&T Network Systems

A S Q C

Q

QUALITY
PRESS

ASQC Quality Press
American Society for Quality Control
310 West Wisconsin Avenue
Milwaukee, Wisconsin 53203

An Approach to Quality Improvement that Works

With an Emphasis on the White-Collar Area

by

A. Donald Stratton
Manager, Quality, AT&T Network Systems

Published by ASQC Quality Press

ISBN 0-87389-038-8

To Lea, Todd, and Mark
with love and gratitude

Preface

This work would not be possible without the contribution of hundreds of Quality Action team members. These fine people come from many varied organizations throughout AT&T as well as some from other companies. It is an extreme privilege to work with these experts in helping define specific ways to improve what they do.

I have been greatly influenced by Dr. W. Edwards Deming's work. I attended his four-day course in October 1983, and took 30 pages of notes to which I still refer to this day. I have read and reread his book *Quality, Productivity, and Competitive Position* as well as its successor *Out of the Crisis*. We owe a lot to NBC for its white paper entitled "If Japan Can, Why Can't We" for helping discover Dr. Deming in the United States.

The study of Dr. Deming's work combined with 30 years of experience provided a unique view of what could be done to implement Dr. Deming's philosophies in a meaningful and efficient way.

I have been using Force Field Analysis successfully since 1970 in various projects. When combined with Cause and Effect studies we have a very powerful tool to develop solutions to problems. It was Dr. Kurt Lewin who developed Force Field Analysis and Dr. Kaoru Ishikawa who developed Cause and Effect.

I have also had the opportunity to work with a great number of good people who have provided help and encouragement along the way. These include Herman Hasselbauer and Jack Pursel who were 20 years ahead of their time in management style, and fine associates such as Billy Terrell, Darrell Storholt, and Carl Perry. My current group, led by people dedicated to the principles of prevention, includes Glenn Kirk, Susan Anderson, Dick Neel, Dick Schuman, Bill Towers, and M. E. Marsh.

My association with Dr. Bikramjit S. Garcha and Ken Charon of Georgia State University as well as Warren Nickel, formerly of IBM, has also been very helpful.

But most of all, the main reason for this book is Dr. Deming. A spark was ignited back in 1983 that still burns brightly. He, too, would be one to light a candle rather than curse the darkness. His encouragement through correspondence has caused an article to be published, many talks and seminars to be given, and now this book.

Contents

Introduction v

PART ONE   The Process 1
  Phase I 1
  Phase II 2
  Phase III 10
  Phase IV 11

PART TWO   The Facilitator 13
  Facilitator Responsibilities 14
  Organizing the Quality Action Team 14
  Key Advice to Facilitators 15

PART THREE   Use of Cause and Effect with Force Field Analysis
Problem Selection...17
  Find the Effect 17
  Force Field Analysis 20
  Cause and Effect/Force Field Analysis
    in a White-Collar Organization 23
  Proving the 85/15 Rule 25
  Long-Term Commitment 25
  Summary 26

PART FOUR   Anatomy of a Successful Field Trip 27
  The Visit — Phase I.......................................27
  The Visit — Phase II......................................30
  The Visit — Phase III 41

PART FIVE   A Case Study 43
  Introduction..43
  Force Field Analysis Examples 45
  Summary 73

Review 75
Closing Comments 81
References 85

Introduction

Let's look at what some current experts and top CEOs say about leadership. Kenneth Blanchard, in *The One Minute Manager,* said:

> The key to good performance is to catch people doing something right.[1]

Michael MacCoby, in *The Leader:*

> Concerning workers, 84 percent believe that workers would work harder if they participated more in managerial decisions.[2]

Robert H. Waterman, co-author of *In Search of Excellence:*

> The underlying message that is most important is that management should have the ability, humility, and will to get out there and to ask the questions and believe the little guy has the ideas.[3]

John Naisbitt, co-author of *Reinventing the Corporation:*

> The top-down authoritarian management style is yielding to a networking style of management where people learn from one another horizontally, where a person gets support and assistance from many different locations. (Also) ... we are shifting from manager as order giver to manager as facilitator.[4]

Donald Petersen, chief executive officer, Ford Motor Company, was quoted in *Quality Progress* magazine:

> We must let go of our traditional systems and unleash the enormous power that people have to bring creative ideas to bear on our problems. Management must establish systems that encourage employee creativity — not stifle it.[5]

James E. Olson, chairman of AT&T and the first National Quality Month chairman, made the following statement in *Quality Progress:*

> ...We must do our best to let *all* people know that we honestly want and need and will *act on* their best thoughts as to how the work they do can be done better.[6]

But the following questions need to be answered:

- How do you get management's attention to act decisively in helping workers do a better job to break down the barriers that prevent quality excellence?
- How do you get workers to participate more in managerial decisions?
- How do you get out there and ask the questions to get the workers' ideas crystallized?
- How do you, as manager, unleash the workers' creative ideas to solve problems?
- And finally, how does management set in motion a way to implement these ideas so that the work can be done better?

Peters and Waterman[3] suggested that one way is MBWA — management by walking around. Not a bad idea. But is it effective? Is management really getting good, solid input by walking around or is there a better way? Let me suggest that there is a better way.

This better way is the use of two disciplines that have been in existence for quite some time. When combined they form a powerful tool to improve quality and productivity. The two disciplines are: *Cause and Effect* (Kaoru Ishikawa) and *Force Field Analysis* (Kurt Lewin). Both disciplines are often referred to, but I know of no one who has combined them as we have. The end result of using CE/FFA (Cause and Effect/Force Field Analysis) lies in well-thought-out management solutions that specify which levels of management need to take action.

We have introduced a process that has helped improve quality in the AT&T Network Operations Organization in Atlanta and many other organizations throughout our company. We are beginning to see success in the use of this process by many AT&T entities and by other companies. Although the process works well in production areas, it is particularly well suited to white-collar areas. This process involves both workers and management and provides each group with a road map of solutions to the problems that inhibit quality excellence.

This process is supported by W. Edwards Deming, who wrote to me and said:

I like your facilitator and force field analysis. Your interest and contribution will do much for American Industry.

I believe we have Dr. Deming's support because our process lays solutions in the lap of management rather than problems. In his book, *Quality, Productivity, and Competitive Position*, Dr. Deming is not enthused about quality circles. He states, "They can accomplish nothing unless the management is prepared to act on suggestions from the QC circle."[7]

This book will walk you through the process from beginning to end. It is the result of working with 75 quality action teams from many diverse organizations, including shop and white-collar functions. The departments include engineering, installation, market planning, legal, customer contracting, public relations, research and development, sales, customer service, telephone centers, and production both inside and outside AT&T. All managers, facilitators, and employees who are part of quality circles will be interested to learn about this powerful technique and how it can be used to improve the quality and productivity of their work. This tool does not replace the many statistical tools that have been developed over the years; however, we believe it will be an additional resource to help businesses achieve excellence.

I would like to address two basic weaknesses of quality circles. Much has been written and discussed about this subject.

The first weakness of a quality circle is that, once formed, management allows it to continue indefinitely. At about the six-month period the quality circle starts planning Sunday picnics or begins organizing bowling leagues — hardly activities that affect the bottom line.

The second weakness of a quality circle is the lack of meaningful management participation. The quality circle often ignores middle management or communicates with the wrong source to get things done. Some companies even empower the facilitator to start action, which recognizes the ineptness of those being paid to get things done in the first place.

The process in this book addresses all of these concerns. It tells you, step-by-step, how to unleash the enormous power people have; how to capitalize on employees' knowledge and skill; how to organize a small group and how to signal an end to its existence after its mission is accomplished. It gives facilitators a powerful tool to use in small group work; it tells how to train facilitators to use this tool; it carefully explains how to involve all levels of management in a productive, nonthreatening way, and describes a technique to encourage participative management which almost all experts agree is the wave of the future for American management.

The entire effort has evolved into a four-phase approach that will be detailed in Part One. These phases are summarized as follows:

Phase I — Getting management's active involvement.

Phase II — Working with a small group on a problem area.

Phase III — Presenting management with solutions to the problem.

Phase IV — Introducing statistical techniques in white-collar areas.

This work is primarily aimed at all management levels but can also be used as a meaningful tool by the blue-collar and white-collar work force. The CE/FFA process can be used by any goods-producing or service industry including government, education, and the private sector.

PART ONE
The Process

Those interested in reading source documents on cause and effect can refer to Kaoru Ishikawa's book, *Guide to Quality Control.*[8] Force field analysis is documented in *Field Theory and Social Science* by Kurt Lewin.[9] The CE/FFA (Cause and Effect/Force Field Analysis) technique allows those closest to the issues to identify problems, their causes, and the forces affecting improvement. It produces straightforward information so management can solve these problems and follow through on areas affected by people, machines, or tools.

Implementation of the CE/FFA approach has evolved into a four-phase process.

Phase I

This involves an intensive session with group management teams that are interested in implementing the quality improvement approach. It is advantageous to work with groups that express interest rather than to force-feed groups with little interest.

The purpose of this session is to help the management team understand its role in implementing and supporting quality improvement. We discuss what management experts say about how to manage as well as what quality experts (i.e., Deming, J. M. Juran, Philip B. Crosby, Armand V. Feigenbaum) say about how to improve quality. When both expert groups say the same thing, management, indeed, has something to listen to and act upon. Many of Deming's points fit this category. Management expert Douglas McGregor,[10] for instance, gave us the X-Y theory of management 25 years ago which Deming discusses, but in a different way. The end result, however, is the same — trust your employees, believe in them, avoid over-management, and drive out fear. Naisbitt stated in his book, *Reinventing the Corporation,*[4] that McGregor was 25 years ahead of his time.

Dr. Juran puts it another way:

> An obstacle to participation by upper managers is their limited experiences and training in managing for quality. They have extensive experience in management of the business and finance but not managing for quality.[11]

We discuss the importance of management forming quality resolution teams (QRT) to act on employee or management group ideas generated by the quality action teams (QAT).

We examine the cause and effect force field studies performed by many different groups. We use examples to emphasize that it is the process that needs fixing and that relatively lower levels of supervision can put these improvements into effect. We don't lose sight of the fact that higher management should get involved in 10 percent of the cases; employees can help in 15 percent of the cases; and middle and lower management levels can act on the remaining 75 percent. This is what I have found in 75 cases.

1

When time permits, we use videotapes to support these points. Deming's videotape, *NBC White Paper*, gives a good overview of what American industry needs to do. Even though it is six years old and addresses only the blue-collar areas, it is still a good tape to gain understanding. We produced our own tapes at AT&T that define the CE/FFA purpose and that show a live session with a group of employees using the CE/FFA process.

In summary, the purpose of Phase I is to get management's commitment to initiate QATs and QRTs and to help them understand the discipline of CE/FFA. This ensures active involvement. Sessions with the management team usually go well if the head person is open minded and willing to lead the team in a new direction. This is so important that I strongly recommend spending time only with groups that have such a leader. Put those who need prodding on the bottom of your priority list; this cannot be overemphasized. If an organization's top person is not totally committed to quality improvement, nothing will happen. I differ from most people on this point — by "top" I do not mean the CEO or president. You can be successful only if the top person is a "location head"; you must have his or her total support for improvement to occur. Usually location heads function independently and have the power to get things done.

Phase II

In Phase II, I work with a small group (six to 10 people) and use the CE/FFA technique to explore a problem. We first go over the basics of the Deming formula and the 85/15 rule. The formula is well known, but few, I find, really understand it. Deming tells us that if quality is improved, productivity is automatically improved. This is done by lowering waste, restarts, and rework. In my work with hundreds of people, I find that everyone, regardless of his or her job, experiences some degree of waste, restarts, or rework. These are the elements that need to be vigorously attacked to improve quality and productivity. Deming says that once this happens higher quality and lower cost goods and services can be provided, enabling you to stay in business and provide more jobs. The next step is to explain the 85/15 rule.

In a recent lecture Deming stated that the common cause/special cause idea originally came from Walter A. Shewhart of the former Bell Telephone Laboratories. Shewhart is renowned for his development of the statistical control chart depicting an in-control or out-of-control process. It is interesting that he, too, gave us the rule of variability.

The numbers are not important. Deming says it is 85/15; Juran says it is 80/20; in my work with 75 teams it comes out to be 82/18. Remarkably close numbers! I reported this at a recent Juran Impro Conference where the audience, which included Juran, appreciated the similar data.

The theory states that defects or problem areas are divided into two distinct parts. The first part is *common causes* — systems or process problems.

These systems or processes can be in production lines or accounting organizations. They are owned and controlled 100 percent by management. Management buys the raw material, schedules the people, machines, or robots to work the material, and determines the precise method and use of every element in the process through customer delivery. Only management can change, fix, or alter the process.

The second part of a defect or problem area is known as *special causes.* These special causes can be attributed to people, machines, or tools and occur approximately 15 percent of the time.

In my 30 years of business experience, when a defect occurred we usually looked for "who" did it. And we usually found someone to take the blame even though that person may have been working in an inefficient process! Management has the absurd need to pin the blame on someone and to be on record as disciplining him or her to ensure that it won't happen again.

Yes, people do make mistakes; and, yes, something must be done about this. If the problem is the process 85 percent of the time — and since management owns the process — then management needs to change the process to get continuous improvement. Management should not blame, coerce, or cajole the person on the job.

Deming demonstrates this succinctly in his red and white bead experiment. I didn't comprehend the experiment until two years after I saw it, but the bead exercise inferred that management must *remove* the red beads from the system — blaming the employee doesn't accomplish anything.

William W. Scherkenbach explains the 85/15 rule well in his book, *The Deming Route to Quality and Productivity.*[12] He makes a valid point. It is wrong for management to give the employee the impression that he or she is responsible only for 15 percent of the problems. The employee may have a good idea of how to fix the process (which is in the 85 percent category) and management should encourage systems to get employee input. This is what CE/FFA does. In the words of one of our customers who asked me to help improve quality in his organization, "CE/FFA is a well-disciplined, powerful, well-organized process for problem solving."

The next step in working with the small group is to have it define the type of quality defects or quality problems it is encountering. It may be that management specifies the problem area to be studied by the small group. This is fine. If the problem is not specified by management, then it becomes the responsibility of a small group of experts. I prefer the latter course. Usually the people closest to the work can come up with an impressive list of problem areas that need exploration.

Once the problems are defined, the facilitator asks the small group to pick the most important one to solve. This whole selection process could take up to an hour.

We call the selected problem the effect; we draw a cause and effect diagram and do a cause and effect study. This process is documented in Ishikawa's book, *Guide to Quality Control.*[8] We stop after each cause is determined, and we explode each cause into its restraining and driving forces, which will be discussed later.

The causes are then ranked in order of importance. This is helpful input for management's consideration. Why tackle cause 10 before causes one through nine are considered? There is one exception to this question. If cause 10 takes one day to implement, and cause one takes four months, then, by all means, tackle cause 10 first and get it out of the way. The prioritization gives management the input to muster the necessary resources for attacking each cause.

We don't stop here, however. It is very important to go to management with suggested solutions — not just problems. This is not something new. Every management book states it, but few people practice it. How many times have you, as manager, heard, "We can't do this because. . ." instead of, "We can do this and here's how. . ." One QAT member, after an 11-hour session, stated, "It was refreshing to focus on solutions rather than just problems as is done in many of these kinds of exercises."

The small group's next step is to do a force field analysis of each cause. This, in essence, is an exercise that helps the small group determine why the cause exists in the first place. This must be known before what to do about it can be determined.

This session usually takes from six to 10 hours, based on the complexity of the problem and accomplishes two objectives: (1) each member becomes trained as a facilitator, and (2) a list of recommendations are formed for action by management.

During facilitator training, the entire six- to 10-hour session allows the leader to get to know each participant in terms of potential as a future facilitator. This is shared with management. Specific training takes place during the session in two ways. I tell the group I'm wearing two hats — a teacher's hat and a facilitator's hat. As a teacher, I relate experiences; as a facilitator, I "do" then ask each member to do.

During the teaching part I share my experiences with the participants which helps move the sessions along. For example: *How to Handle the Introduction*. I show an 18-minute tape produced at AT&T that gives a brief introduction to CE/FFA and shows a live group using the process with appropriate verbal communication. After the tape, I describe the Deming formula and discuss it in relation to the 85/15 rule. Two purposes are accomplished in this introduction: (1) the group loosens up, and (2) I gain credibility. The group begins to realize this is not another witch hunt. One conferee's comment sticks in my mind: "I thought this was just going to be another quality meeting where someone gets beat over the head. Instead I became convinced management really wanted to know how to make things better."

Tools to Use. I like to use the 18-minute videotape entitled *Cause and Effect/Force Field Analysis*, a television and a videocassette player to show the tape, and a booklet entitled *Quality Teams — How Do They Work.* The

4

booklet is helpful for management and participants of QATs and QRTs and can be examined prior to a CE/FFA session. A 54 inch by 48 inch wide easel or two small easels put together and a roll of masking tape are also needed.

How to Set Up the Room. For the CE/FFA session I like a room big enough for six to 10 participants to sit in a semicircle around the easel. Tables are not necessary and, in fact, hinder good discussion. However, we place enough tables behind the semicircle to accommodate six observers. Allowing observers works provided the facilitator enforces strict rules of nonparticipation. Their questions can be handled at break or during lunch.

Minimum/Maximum Number of Participants. Much has been written about the ideal size of a small group. I find that less than six people or more than 10 people tends to be ineffective. In the group of six or less there is usually not enough experience; however, I have worked with groups as small as five and as large as 14 and everything worked out — but I'm sure luck was on my side.

How to Memorize Names of Participants. During the CE/FFA videotape I pass around a sign-up sheet that reads "First — Last — Name." It's surprising how many people use initials if you don't specify this. While the tape is playing, I memorize the names of each group member. This may be a bit manipulative, but I find people really appreciate being called by name. I do admit to them later how I did it so they may also use this technique.

How to Handle the Verbose or Quiet Participant. In both situations the group usually curbs too much participation or bolsters someone who may not feel good about participating. As facilitator, one must break in if the verbose participant is off on a tangent or out of control; I believe the quiet participant should be left alone. Generally the group will take care of this, although I have seen some people not say one word during their entire time together. This could be due to inexperience or some other unknown problem. However, I do check with management to assure the session will not place undue pressure on any individual.

How to Handle Disagreement. Healthy disagreement sometimes leads to a breakthrough in understanding a particular problem. Compromise can develop to strengthen the situation. The key word here is "healthy." This is generally the case; the basis for disagreement is usually someone's lack of knowledge or understanding about a particular situation. It is good to try to discuss as many views as possible, and reach some kind of consensus.

The Cause and Effect Basics. This is easy to explain by using the cause and effect diagram (see page 18). I recently was reminded of the process by a friend while we discussed cholesterol. Cholesterol is the effect. The doctor helps the patient understand the causes, inappropriate diet usually being the number one cause. In some people, however, chemical imbalance is the primary cause, not diet. In this case the exact cause must be determined before an appropriate "fix" can be made. There may be many causes in business-related examples — improper grounding, late payments, or inaccurate office records. I have seen as few as five causes and as many as 12 causes developed by small groups.

The Force Field Analysis Basics. The use of the force field diagram is the best way to explain this process (see page 20). Let's discuss an example. Suppose one of the causes is lack of training. First, we write the words *The level of* on top of the easel pad and then use words to finish the sentence as best as possible. This may then read *The level of training is too low.* Another example could be *The level of absenteeism is too high.* The idea is to define the cause in terms of the level being too high or too low.

After the *cause* is defined in terms of level (too high or too low), draw a horizontal line across the middle of the page. Ask the group to picture the level of training being at the level of the horizontal line. Then say if the level of training is too low, something is keeping that horizontal line where it is — we call these *restraining forces.*

Write the words *restraining forces* on the top of the easel paper. The group then defines the restraining forces. Examples in this case could be: *too much on-the-job training by inexperienced people, lack of manuals, lack of dedicated training space, and lack of trainers.* When all known restraining forces are listed across the page, we go to the bottom of the page and ask what can be done about each restraining force and write down the answers. We call these *driving forces* and write these words on the bottom of the page. Draw an arrow to the horizontal line to meet the arrow coming from the restraining force. Then ask what other restraining forces this particular driving force would affect and draw lines to the corresponding restraining force. (Refer to the example on page 20.) The whole idea of force field will become more apparent as you read through the examples in *Part Three, Use of Cause and Effect Together.*

The Importance of Defining Solutions. As mentioned in the introduction, it is important to present management with suggested solutions rather than just problems. Employees closest to the work often have very good ideas about what these solutions are and can make solid recommendations. If we stop at the problem definition as presented in the classic cause and effect, we miss the opportunity for needed input.

Why it is Important for Each Participant to "Do" a Force Field Analysis. The prime facilitator, trained and certified in CE/FFA, does the introduction, the cause and effect analysis, and the first force field analysis. The facilitator then "passes the crayon" to the work-group member sitting in the first chair on either side of the semicircle. The reasons for this are:

- The group tries to understand the process, knowing they will have to do it.
- This is the best way to learn the process.
- It gives the facilitator a good idea of who can excel at this; this is shared with management.

6

- It guarantees high-level group involvement.
- It helps avoid facilitator burnout during the 10-hour period.

These reasons answer a question that often comes up: "You're so good at this, why don't you do it all?" Through extensive experience with this, I can report that something very positive happens when a group member gets up and participates — the entire group benefits. I highly recommend this method.

The Role of the Facilitator While the Group Member Does the FFA. The facilitator sits in the chair of the group member doing the FFA and participates in the roles of gatekeeper, encourager, harmonizer, consensus seeker, feedback seeker, standard setter, and processor. Let's examine each role. (Further information may be obtained by reading *Making Meetings Work* by Leland P. Bradford.[13])

- *Gatekeeper* — Keeps the "door open" for timid, less talkative members to contribute.
- *Encourager* — Shows in a caring way how members of the group can help one another.
- *Harmonizer* — Helps in a conflict situation; gets heavily involved when the group polarizes and cannot reach a consensus.
- *Consensus seeker* — Gets members to reach agreement on what is best for all; keeps discussion on target.
- *Feedback seeker* — Gives individual and group feedback on progress. Feedback may be appropriate during the session, break, lunch, or dinner.
- *Standard setter* — Presents the image of being ready for work, on time, interested in the group's progress, and participating where appropriate during an intense six to 10 hours. The best compliment is the group wanting to work harder than the facilitator, opting to work late or come in early once they are comfortable with the process.
- *Processor* — Keeps track of using the CE/FFA process.

The presentation time frame is usually as follows:

- *Introduction* — One-half hour, including the 18-minute videotape.
- *Cause and effect* — 45 minutes to one and one-half hours.
- *The first four force fields* — 45 minutes to one hour each.
- *Remaining force fields* — 20 to 30 minutes each.
- *Summary* — One-half hour.

With these time frames in mind the facilitator can closely monitor the process time length.

Monitoring the content of the process is quite different and has already been mentioned.

The Importance of Rank Ordering Causes. It is important to prioritize causes for management to understand what employees view as important problems.

We recommend tackling the most difficult problems first. As mentioned, if one cause would take hours to solve and another months, eliminate the lesser problem first.

How to Summarize. When all force fields are completed the facilitator again takes over. We tape up and display the original cause and effect analysis and all the force field analyses. The facilitator starts with cause number one and asks the group whether the first driving force is the responsibility of management or the worker to implement. When all driving forces are tallied this gives a rough estimate of how many are management or process problems and how many are worker, machine, or tool problems (see Phase III discussion on page 10).

The facilitator also asks group members what level of supervision, in their view, should be responsible for implementing each driving force. These data are recorded and summarized.

How to Organize the Management Presentation. It is recommended that the senior-level manager open this session and set the tone. He or she should restate the session's purpose and thank the QAT.

The facilitator then reviews the basic Deming theory and the 85/15 rule. Next, the tally sheet showing the number of management- or worker-controlled driving forces is discussed.

At this point the facilitator introduces the group member who will explain the cause and effect analysis. You will recall that the work group started to actually "do" force field analysis number two. The facilitator did the cause and effect and the first force field analysis. Prior to the Phase III session, the facilitator asks for volunteers to explain the cause and effect and first force field analysis, excluding those who already have a force field to explain. This usually works smoothly and gives almost everyone a chance to participate in the Phase III session.

Lately, we have worked it out so everyone makes a presentation. One person can be assigned restraining forces and another driving forces of the same force field. This gives everyone a shot and makes for a better session.

At the end of the Phase III session the facilitator sums up by going to the board and writing the words *What's next.* The suggestion is made to form a QRT made up of supervisors, with representation from the QAT. Two representatives are sufficient to provide the continuity necessary for understanding the analysis. A chairman should be appointed at this stage. At the very end it is good for the facilitator to recognize the hard work of the QAT. This can be done by verbal praise and by presenting a certificate to each participant.

How to Condition Management for Phase III. It is important to condition management prior to the Phase III session. Recommendations are going to be made that may challenge or embarrass some of the attendees, i.e., "Why didn't I think of that?" or "Why didn't this get fixed five years ago!" The conditioning should aim at eliminating management's defensiveness or brow-beating.

When opening the Phase III session it is a good idea for the facilitator to announce that it is an input session, *not* an action determination session. This will happen later in the QRT.

How to Track Results. The best way for a QRT to track results is to *not* issue minutes. It is difficult, at best, to find any action items in most sets of minutes. Volumes are often written but seldom does one find concise descriptions of the action item, the responsible person, or the target dates — much less anything about estimated annual cost savings.

I recommend, therefore, using a simple spread sheet with the following headings typed across the page:

- *Action item* (called driving force).
- *Person responsible.*
- *Target date.*
- *Actual date.*
- *Comments.*
- *Estimated annual cost savings.*

This document should become known as the action register list of each QRT and each quality review team. It should be used to track progress and determine if higher management levels should get involved.

How to Have Fun Doing the CE/FFA. Six to 10 hours may not seem like a long time. Use of CE/FFA demands a high level of group attention and participation. When levity can be utilized, it is certainly encouraged. This relieves tension and stress and often aids the group along the way. The form of levity should be relating group experiences, not by telling jokes. It is amazing how a group member's comment or action can lead to such humor.

The best way facilitators learn is by *doing* it. This is why we give as many people as possible the opportunity to actually lead the force field analysis session.

We found that it is also possible to include a small group of observers. These people are separated from the work group by space or tables and are not allowed to participate in the actual work group session. They are asked to write down their questions and discuss them with the experienced facilitator during break or in a special session.

The first reaction of most people is to reject the idea of observers. My experience indicates it can work if the facilitator strictly enforces the above criteria. The workers and observers both respond beautifully when they know the rules!

We call these small groups QATs.

Phase III

This is a most exciting event. The workers get a chance to share their findings with the management team — and I have seen six management levels together at the same time. Good quality presentations result in spite of lack of extensive preparation or written reports.

The QAT actually uses the same easel charts it used in the work session. As we discussed, at least one participant always writes everything down in the same format on 8 inch by 11 inch paper, so this can be reproduced and handed out at the session with management.

During a particular Phase III session, a manager leaned over during the middle of an employee presentation and whispered to me, "It took me six months at 'charm school' to learn how to talk with such authority. How did you do this in 10 hours!"

It's not teaching people how talk — it's giving them a tool that allows them to express information they know better than anyone else. I have seen hundreds of people do this with a zero percent failure rate — and this covers a wide spectrum of talent, from labor grades in installation to software developers with PhDs. The common thread is their individual job expertise. Yes, this is perhaps the most underutilized resource in industry today. All we have to do is turn it on. As Tom Melohn, president of North American Tool and Die, recently said, "They're out there, guys — go out and find 'em."

At this point it is good for the facilitator to recognize the good work of the small group. Very often these people have never made a presentation to their own supervisor let alone four levels of management. This recognition can be verbal praise and the presentation of certificates to each group member.

After the employee presentations and while the entire group, workers, and management are still present, it is important for the facilitator to emphasize the summary results. This could be done at the very beginning of the session, depending on the situation as the facilitator sees it.

Results to the following data:

- 63 Driving forces (solutions).
- 82 percent — Common causes (management-owned process problems).
- 18 percent — Special causes (people, machines, and tools).

Of the 82 percent management-controlled problems:

- 60 percent — Can be implemented at lower levels of management.
- 30 percent — Can be implemented by middle management.
- 10 percent — Can be implemented by higher management.

This next part is designed to address Deming's concern about management's lack of support for employee groups. I actually get management's commitment to "do something"; as mentioned in the Phase I discussion, we pin down management's responsibility.

I ask management to form a QRT that will address each driving force. As mentioned before, I ask them to use a spread sheet to track progress and send it to me. The spread sheet is nothing new but, used in this context, becomes a powerful communication tool with the original QAT.

I also recommend to the highest level person that he or she chair a quality review team that will meet at least once every other month to review progress. The department heads then chair their own QRT; these heads make up the review team.

Phase IV

In this phase we take a micro view of the process. Every job, whether it's production or white collar, is part of a process. The purpose of this examination is to pinpoint measurable areas of the process, build a data base in these measurements, identify potential or real problem areas, and take action to correct and prevent future problems.

Flow charts are a good method to use for identifying the overall process and pinpointing the areas requiring specific measurement. It is possible to use any one of the statistical tools to understand the data, but we recommend the use of control charts. AT&T's *Statistical Quality Control Handbook*[14] is an excellent source document that explains the various statistical methods.

As discussed in the handbook, we encourage using the triangle concept of a supervisor, a statistical analyst, and a key person in the group being analyzed. This small group (1) meets periodically to go over the data, (2) determines its meaning, (3) decides on action to be taken as a result of the data, or (4) seeks more data.

It is now important to understand that this entire four-phase approach is as appropriate for white-collar jobs as it is for production-oriented jobs. In fact, it is critical to immediately implement these phases in the white-collar areas.

PART TWO
The Facilitator

Let's now go back to the Phase II area and discuss the facilitator's role — organization, problem selection, and cause and effect and force field analysis. The CE/FFA technique uses small groups drawn from either manufacturing or service and administrative operations. The team works with a trained, experienced facilitator knowledgeable about both cause and effect and force field analysis.

All the reading and looking at videotapes, however, cannot replace first-hand experience. We have found that the only way to train facilitators is to put them in a "learn by doing" environment. It reminds me of a Chinese proverb used by Kip Rogers, ASQC human resources chairman, at the 40th Annual Quality Congress:

"I hear, I forget
I see, I remember
I do, I understand."

This is extremely appropriate regarding facilitator training. When they do it, they understand it!

Learning by doing training takes from six to 10 hours. Trainees are part of a small group (six to 10 people) assigned to work on a specific quality problem using the CE/FFA technique. They observe an experienced facilitator develop the problem area to be explored using group consensus, perform a modified cause and effect study on the major problem selected, and initiate a force field analysis on the first cause.

Under the careful eye of an experienced facilitator, the facilitator trainees then initiate a force field analysis on the remaining causes. In addition to this hands-on experience, we provide new facilitators with the following ongoing support:

- *Access to an experienced facilitator* — Helps the new person by answering questions prior to working with the first group and provides ongoing support.
- *Observation of working groups* — Works well as long as the observers are separated from the group and are not allowed to participate in discussions. Questions are written down and answered by an experienced facilitator during break, after the completed CE/FFA session, or in a post-session facilitator and observer discussion.
- *Workbook describing facilitator role and CE/FFA process* — Pamphlet-size booklets describing the process with illustrations; this is aimed primarily at the work force but is also a good summary for the facilitator.
- *An 18-minute videotape* — Shows a live group using the CE/FFA process.
- *Direct contact with the process initiator* — Includes private sessions prior to the actual small group session or consultation during a small group session.

- *Periodic facilitator seminars* — Reviews the basics of how to get started and what the cause and effect and force field analysis is; discusses unique problems the facilitators have encountered since the last session. New ideas are also introduced.

The facilitator becomes the key person to provide solid communication between workers and management — defining what needs to be done for process improvement.

Let's examine the facilitator's role more closely.

Facilitator Responsibilities

The facilitator must create an atmosphere that allows free and open group discussions. The facilitator must avoid the temptation to make speeches, direct the discussion, or squelch ideas inappropriate to the topic at hand. Sometimes this may mean letting one to two participants ramble on rather than risk losing credibility for the discussion openness. There is a time, however, when "war stories" must be labeled as such and limited. You can use the cause and effect and force field analysis structure to your advantage in leading group discussion and keeping on track.

All discussions should be free of any type of facilitator censorship. Without understanding this at the beginning, some of the best ideas could be lost before they are even mentioned. This does not mean that the process becomes a gripe session or free-for-all. Rather, the facilitator has to keep the group on track without stifling the contributions of its members.

As facilitator, your opinions are not appropriate; you are seeking the group's ideas. On the other hand, your job is to help clarify some of the group's thoughts. Do this by suggesting examples to reduce confusion. It is also appropriate to suggest combining similar thoughts or ideas, with the permission of those who raised them. And when the group seems to meander, you must get them back on track.

You also have the responsibility to record on paper as accurately and concisely as possible the group's thoughts.

Exercising some of these responsibilities requires practice. Leadership talent is important; in fact, one of the reasons a facilitator is selected is because of his or her ability to lead small group discussions.

Organizing the Quality Action Team

The group applying these techniques should include six to 10 people who share common work problems. Although any group can use these techniques, it is best to involve employees closest to the problems who are capable of identifying what actually happens on the job.

Each location and organization uses QATs in its own special way, but may give it different names. You should be able to adapt these techniques to the objectives of your version of the QAT.

Formal quality improvement groups operate as part of a larger quality program undertaken by the location or organization. They usually meet at regular intervals and have follow-up responsibilities for pursuing problems to positive resolutions. These techniques can help groups better identify problems and pinpoint action to be taken.

Ad hoc groups are usually organized by supervisors trying to come to grips with a specific problem. These groups continue to function until the specific need has been satisfied. Although *ad hoc* groups can try to immediately tackle their problem, they should consider more than just the problem at hand to take full advantage of these techniques.

These techniques have been used successfully with workers at all levels as well as with many levels of management. They have been used successfully combining workers and supervisors in a group.

The group should schedule six to 10 hours to complete one exercise. This time can be broken into smaller units, but to be productive we recommend meeting for at least two consecutive hours. Management approval is obviously necessary, if only to authorize the time and to provide the meeting place.

The ideal situation allows the group to meet for two sessions in two days. The first day would require six consecutive hours of work; the second day would require two to six more hours, depending on the problem. In this way, the group stays intact and on target.

However, this ideal situation is not always attainable. We have seen groups meet once weekly for one-hour sessions and still get the job done. A strong facilitator is required especially in these situations.

Key Advice to Facilitators

At the conclusion of a recent session the top executive, who observed all three phases, made this statement:

> You obviously have acquired some good facilitator skills. Can you share with us what you feel contributes to being a good facilitator?

My answer was:

> There is one thing that stands out as being the most important. You must sincerely believe that the group you are working with truly knows more about the problem than you do. When you believe this the rest falls into place.

All that you have read, so far, about the facilitator skills are important. The answer above is the key advice that will carry any facilitator through. If you truly believe that the small group's collective knowledge is far superior to any one person's knowledge — including your own — then you will succeed by using your own common sense to get that knowledge out and be able to capitalize on it.

You just do not say to yourself that now is the time to act as gatekeeper, or harmonizer, etc. All this should fall into place naturally. It is somewhat similar to an athlete. He or she practices on the elements profusely but during the game what one learned in practice must flow naturally. Your work as a facilitator works the same way.

Another bit of advice is that it's really best if you, as facilitator, are not an expert on the problem being explored. I recall working with another company in the area of accounting. I was not an expert in accounting or familiar with the company. A small group member made quite a plea that he should only work with groups in accounting. He went on for 10 minutes building his case. When he finished I made the following comment:

> Bill, I don't know anything about accounting or this company for that matter. How am I doing as facilitator?

The group laughed, Bill blushed, and we carried on.

PART THREE
Use of Cause and Effect with
Force Field Analysis Problem Selection

Selection of the problem for analysis is an important step. This is why we suggest that every group start with this procedure, even if chartered to pursue a specific issue. In addition to focusing on a specific quality problem area, the selection process also helps to establish the tone of the meeting and the level of give and take. However, it is perfectly fine for management to select the problem.

Find the Effect

Start the selection process by drawing the broken vertical line that separates the major causes from the effect in the Ishikawa chart (Figure 1). Ask the group for quality problems that are important to them and write out the list on the right-hand side of the chart.

Figure 1

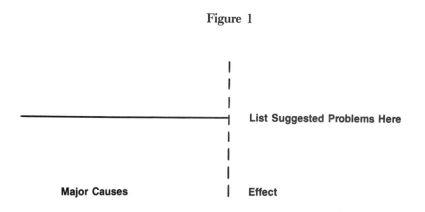

Effects are those quality problems or defects that occur on the specific job to which each group is assigned. In a production environment these could be the kinds of defects for which the individual workers or work unit might be written up in a quality audit. In white-collar groups you will find that groups often define causes when asked to define effects. (See *The Visit — Phase II*, page 30.)

One of the difficulties in getting started is distinguishing between the effects and the major causes. Often participants will want to suggest problems that occurred earlier in the work flow, before reaching them, such as receiving wrong or faulty components. Although that may be a serious problem, the group should focus on the results of having to work with those kinds of situations — results for which they are likely to be criticized.

Note each of these effects on the diagram. When several suggestions are similar, ask the individuals if they might be talking about the same thing. Don't make assumptions. It's important to let the group members clarify for themselves what they really mean.

After a reasonable time — approximately 30 to 45 minutes — or when the suggestions seem exhausted, have the group reach a consensus on the *one effect* that is the most significant problem affecting quality.

Start by letting group members volunteer their choices and reasons for these choices. If the consensus does not appear to be forthcoming, poll the group. Work on what appears to be a clear majority of opinion, even if you have to poll the group several times to narrow the selection. It's important that all group members own the problem.

There are times when management may wish to specify the quality effect being examined. This may be the result of negative trends, quality audits, or anticipation of quality-type problems in specific areas. The key here is to put the correct group of people together to examine that problem.

Figure 2

Cause and Effect Study

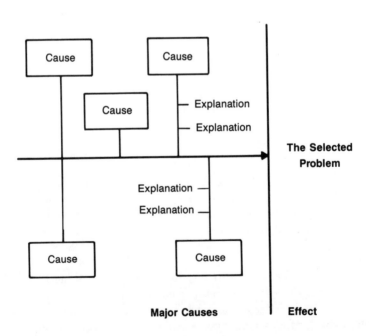

After the group determines the number one quality problem, the facilitator draws a cause and effect diagram on the pad and asks them to suggest its major causes. The facilitator diagrams those causes, shown in Figure 2. Right now the group looks for causes only; the process calls for identifying solutions later, during the force field analysis. If someone does volunteer a solution, the facilitator should ask that person to write it down so that it won't be forgotten; the group then keeps looking for major causes of the effect.

At this point, the group must organize the causes into order of importance. Figure 3 shows a completed cause and effect study. This particular group started with 10 possible causes; by setting priorities and reorganizing, they narrowed the list to six.

Figure 3

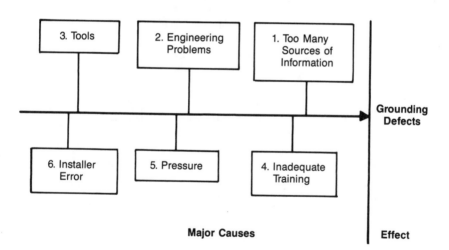

GROUNDING DEFECTS

Major Causes   Effect

Force Field Analysis

This concept states that any problem or situation is a result of forces acting upon it. Restraining forces keep the problem situation at its current level. Driving forces push the situation towards improvement. Restraining forces are the causes of problems; driving forces are solutions.

In a force field analysis, a horizontal line is used to represent the current problem level (Figure 4). We described the current force field situation by stating whether the level of the problem is too high or too low. In this case, for example, the level of information (cause 1) is too high.

Figure 4

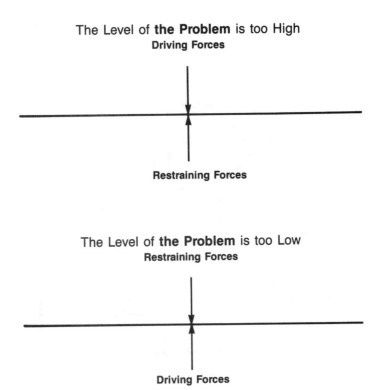

The Level of **the Problem** is too High
Driving Forces

Restraining Forces

The Level of **the Problem** is too Low
Restraining Forces

Driving Forces

Figure 4 shows how a problem is diagrammed in a force field analysis. If the problem situation is too high, restraining forces are shown as pushing it up while driving forces drive it down towards improvement. If the problem situation is too low, restraining forces push it down, and driving forces push it up towards improvement.

The force field analysis process begins by the group examining the most important cause identified in the cause and effect study and states whether its level is too high or too low. The group then identifies the restraining forces — the situations and events that keep the problem situation at its current level. In the group we have been referring to, the first analysis was "the level of information is too high." This group of installers and supervisors determined that too many sources of information was the number one cause of grounding nonconformances before verification and correction.

They saw the following restraining forces contributing to that problem level:

- Different grounding requirements for each type of equipment.
- Changes in requirements and timely notification.
- Different interpretations of grounding requirements.
- Workers' inability to absorb too many information sources, compounded by the need for additional clerical support and the need to distribute information over a wide geographic area.

Each of these restraining forces is diagrammed in Figure 5. For forces that compound the restraining force, the vectors are connected to the restraining force itself.

Figure 5

The Level of **Information** is too High

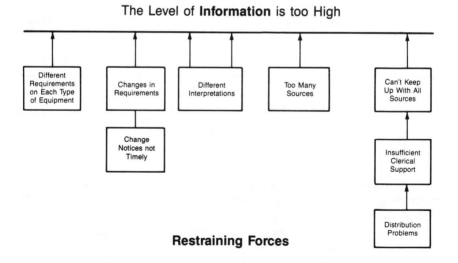

Restraining Forces

As the group identifies restraining forces, someone is likely to offer solutions. Just as when solutions were offered during the cause and effect study, the facilitator should acknowledge the value of the idea, remind the group that it is looking for restraining forces only at this time, and ask the person to make a note of the solution.

After the group identifies a significant number of restraining forces, it starts looking for driving forces to counter each specific restraining force (Figure 6). In this case, the group identified "clarify information" and "make it simple" as driving forces to counter "different interpretations."

Figure 6

The Level of **Information** is too High

Driving Forces

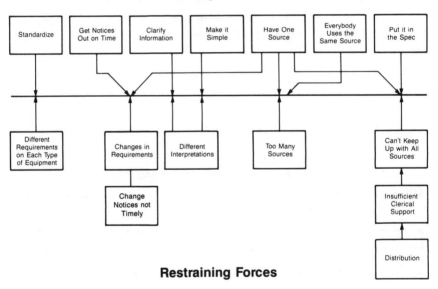

Restraining Forces

In some cases, more than one driving force will be found for a specific restraining force. The diagram shows this by pointing the arrows from both driving forces towards the opposing restraining forces.

Often, one driving force will counter more than one restraining force. Again, arrows from that driving force are shown in opposition to all the restraining forces it can counter. The example has "have one source of information" as a driving force countering "changes in requirements," "too many sources," and "can't keep up with all sources."

22

Once the group identifies all the restraining forces and their opposing driving forces for a specific major cause of a quality problem, it has completed a force field analysis. The diagram presents a specific problem along with a prescription for its solution. To complete the process, the group goes through each major cause identified in the cause and effect study and performs the same kind of force field analysis.

It should be noted that this process is self-correcting. In this case, the group initially identified six causes, as shown in Figure 5. As the group was pursuing the force field analysis, however, it discovered a seventh cause, poor motivation. This cause was then ranked third in order of importance.

The analysis of low motivation shows how the cause and effect study/force field analysis process is self-correcting. Low motivation was first suggested as a possible restraining force that kept the level of information too high (i.e., as a contributing factor to one of the six original causes). The person suggesting low motivation was hesitant to raise the issue, but the group's openness eventually created an atmosphere in which he felt free to address it. Once the issue was out in the open, it quickly became apparent to the group that low motivation did not really contribute to the level of information being too high. They saw that it belonged with the major causes of grounding problems.

Cause and Effect/Force Field Analysis in a White-Collar Organization

Let's look at another example of CE/FFA in action — this time in a service setting attempting to reduce late payments. This group started with a list of 12 possible causes, then narrowed the list down to eight major causes, as shown in Figure 7. These white-collar workers selected "inadequate training" as the number one cause of late payments in their organization; that is, the group determined that "the level of training is too low."

Figure 7

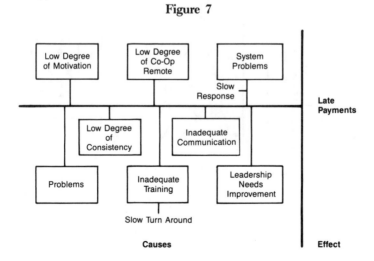

23

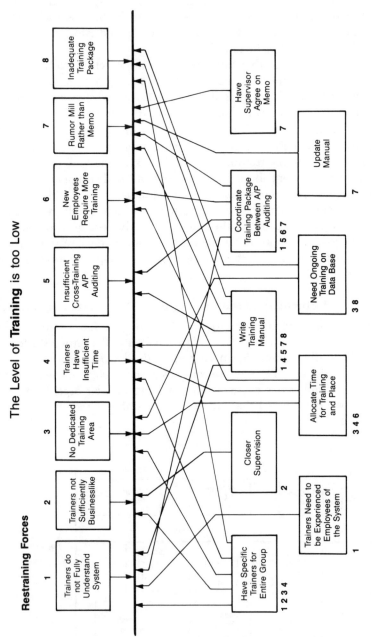

Figure 8

The Level of **Training** is too Low

The group identified the following restraining forces:

- Trainers were not sufficiently businesslike.
- There was no specific training area.
- Trainers had insufficient time.
- There was insufficient cross-training of accounts payable and auditing personnel.
- New employees required more training.
- Information was communicated through rumor rather than memorandum.
- The training package was inadequate.

The resulting CE/FFA diagram is shown in Figure 8.

The arrow crossings can make the CE/FFA analysis results difficult to understand at first, especially if the arrows are all the same color. To make the results clearer, the restraining forces can be numbered, and the driving forces assigned numbers according to the restraining force they affect, as was done in Figure 8. The numbers can be used to help understand which driving forces affect which restraining forces.

Proving the 85/15 Rule

The completed force field analysis diagram provides a significant list of positive suggestions to solving quality problems. To determine ownership of the problems, the group writes "M" for management or "W" for worker on each driving force. (When you add up the totals, you will find that management owns close to 85 percent of the driving forces, just as Deming and Juran suggest.) Next, the facilitator asks the group to specify the supervision level necessary to act on each driving force, and writes these levels next to each "M." This is the group's best guess, of course, but experience shows these groups are usually quite accurate in their assessment. As mentioned, we have surprisingly found that a great number of solutions can be put in place by lower levels — 60 percent can be resolved at first and second levels of management.

The same process is repeated for worker-related solutions. Communicating these ideas to workers becomes the responsibility of both group and management.

The group's findings should be documented. This means copying the developed diagrams as part of the cause and effect study and each force field analysis. Documentation should also include notes and pertinent observations. In preparing these documents, time is of the essence and elegance does not matter. This documentation can be shared with all managers who attend the Phase III session; it can also be used by the QRT.

Long-Term Commitment

Some of the issues identified by the groups are products of years -— perhaps even decades — of doing things one way. Changing traditional patterns is always difficult. In fact, Deming suggests that such basic changes happen slowly, over a period of years. Quality is both a start and commitment. The quality resolution team will require reasonable time to pursue these issues to appropriate resolutions for improvement to result.

Summary

In the first phase we discuss the following with the management team location head:

- Get active involvement in a quality improvement program.
- Explain the factors of a good quality improvement program.
- Explain the Deming philosophy and key points of other quality philosophies.
- Discuss what other companies are doing.
- Explain CE/FFA.

In Phase II we actually conduct a CE/FFA session with six to 10 people. The purpose is twofold:

- Develop recommendations to make the process better, using the CE/FFA technique.
- Train participants and observers to be facilitators, using the CE/FFA technique.

In Phase III the small group makes a presentation to management on their findings. This purpose is also twofold:

- Help management understand employees' views of what needs to be done to help them do their work better.
- Get management commitment to establish QRTs to address and resolve problems.

In actual work with 75 QATs and the resulting QRTs, we find this process to work, provided management is 100 percent supportive.

Phase IV is primarily aimed at white-collar groups where in-process checks are instituted to ensure quality. Flow charts depicting specific areas to be measured are introduced and data are gathered at these checkpoints. The data are analyzed to determine if the process is in control and how it can be continuously improved.

PART FOUR
Anatomy of a Successful Field Trip

The initial contact:

"Hello, Don, this is Jim. Can you come and help us improve the quality of our operation? I've been working with a group in the planning area and, as you know, I'm the quality coordinator for our organization."

"Sure, Jim. I'd be glad to help. Our only problem will be getting it on the calendar. What do you have in mind?"

"Well, we are familiar with your four-phase approach and believe that the first three phases are appropriate right now. In fact the manager volunteered his group at our last staff meeting. Both heard you talk and want CE/FFA implemented in our organization."

"Are you saying you want the entire three-phase approach?"

"Yes."

"Who will Phase I be directed at?"

"The entire manager's organization."

"I'll give the Phase I discussion to the entire group, then I will ask the management team to leave, and I'll work with the small group using the CE/FFA technique. How much time would you like to have for the Phase I session? What do you want to accomplish?"

"Well, many of these people have already heard about your work but they haven't seen you face-to-face. So, we would really like you to give them a basic appreciation of your philosophy, whatever results you can share, and, of course, your CE/FFA approach and how that works."

"To accomplish this, I can do anything from one hour to four hours, based on the time you have. How about a two-hour session with the entire group? Would you want me to include the 18-minute videotape on CE/FFA showing a live group using the disciplines?"

"Absolutely. Let's plan on the two-hour session including the tape."

"Sounds to me like we'll need two days. First I'll give the Phase I talk, then the supervisors can leave, and I'll begin work with the QAT. This will take from six to 10 hours and, as you know, we refer to this as Phase II. Then we will begin Phase III, which will last from one to two hours."

"Sounds good. What will you need?"

"For Phase I, a 35 mm carousel projector with a remote control that reaches the screen. I'll also need a videocassette player and an easel to write on — any size. For Phase II, the most important thing is a 54 inch by 48 inch easel. If you can't find one you can put two small easels together. We won't need any other equipment for Phase III.

"Thanks, Don. We're all looking forward to this."

The Visit — Phase I

I arrived at the site of the Phase I session at 8:30 am. The room was magnificent. It was a small, plush theater-style room with a stage, large screen, and all kinds of front screen audiovisual equipment.

The audience was also impressive. It was obvious they were high level, technically oriented people who knew what they were about.

The talk went well. No talk is ever exactly the same. Many times I find myself reacting to even a positive or negative gesture and inserting a comment that wasn't thought of previously. In this particular talk it appeared the group was seeking more definitions of Deming's "drive out fear" point.

I shared with them an experience I encountered at the Juran Institute Impro Program. I was a speaker and arranged a professional quality display. I believe it was the first for AT&T. After my presentation, I went to the display area where many conferees were gathered.

One of the visuals in the display was a list of principles we developed, one of which concentrated on driving out fear. A comment was made by a young conferee: "Why would anyone include the idea that fear even exists in our company?" I tried to explain to her that fear manifests itself in many different ways. I told her about a QAT I worked with that constantly referred to its boss's office as "never-never land." All 10 QAT members were unanimous in their opinion that the boss's office was not a place to be seen or heard. It was a place to be avoided, at all costs. I asked the conferee how long she worked for her company. She replied, "Two years." I just said, "Keep your eyes open; you'll find fear. It's still there unfortunately."

Because they seemed tuned in to this particular point I shared with this group the ideas of William W. Scherkenbach of Ford Motor Company on this subject.[12]

Scherkenbach refers to the thousands of hours of research that can result from a CEO's raised eyebrow at a meeting. My experience says the raised eyebrow can come from someone at a level a lot lower than the CEO. How many hours of work are wasted chasing rainbows, finding answers to questions that aren't ever asked — because of fear?

Scherkenbach also refers to the wasted hours putting together black books for people who go to meetings so they'll know all the answers. The reason people can't go to meetings without all the answers is fear. We ought to be able to say, "I'll find out in 24 hours," rather than lose three man-weeks putting the bogus black book together. I agree.

The other Deming point that many people challenge is the elimination of numerical goals and work standards. This group was no exception. Some say, "You only get what you measure." I say, "And that's all you get!" Time and time again I have talked to people in many companies who could do more than they were asked to do. This, of course, is after experience sets in — not during the initial stages that require maximum training. I always ask this question of the measurement advocates: "If you're so enamored with measuring people why don't you tell them you'll let them go home when they meet their bogie?" Most people would be home by 2:00 pm in a normal work day. You see — and here is the enormous waste — we certainly wouldn't let them go home when they meet their bogie, so they really are doing what management asks: five hours of work in eight hours. What a waste.

The other idea stressed by Deming is how we use numbers to manage people. He strongly advocates the elimination of slogans, exhortations, and acclamation aimed at the worker to shape up. We had a situation where someone had the idea of motivating engineers to do a better job by drawing cartoons of engineers making mistakes. When management began to understand that the *process* the engineers were working on was the problem, not the engineers, the cartoons disappeared almost overnight.

The best example of the negative effect numbers can have on the work force is an experience with an accounting group. In working with the QAT, one cause of late payments was low morale in the group. A fairly low level employee was at the easel doing the force field analysis on low morale. One of the group members said, "Let's face it. What causes low morale in our group are the charts."

This was a group of 140 employees divided into four sections. The group was measured by four critical measures each month. The head supervisor ranked each section and hung four large charts each month so everyone would know who was first, second, third, and fourth. God help the group that came in last! They were ostracized and supervisors fought verbally in full view of the work force. Group 1 would not eat lunch with groups 2, 3, and 4, etc.

Then came the Phase III session with management. The head supervisor was a 240-pound crewcut type, with a manufacturing background. He wanted to tear me apart during the Phase I discussion on eliminating work standards. During the Phase II session, though, he was remarkably interested, sensitive, and attentive while hearing his own people discuss how to make things better.

When the young employee got up to discuss why morale was low, the entire QAT group gasped at the thought of little Nancy announcing to the management team the idea of taking down the charts. When she did, Mr. Boss just wrote it down. After the meeting he went out on the floor and took the charts down himself. He became an overnight hero; his results improved remarkably. In six months, late payments rose from 95 percent to 97 percent on time. Sounds like a small improvement. When looking at this improvement from the customer's viewpoint, however, there were about 1,000 more satisfied customers at the 97 percent level than at the 95 percent level. This is a remarkable improvement in results.

It's not that management should throw away all data. They need data to know what's going on. The important thing to gain from this experience is how to use the data when involving employees.

The group seemed to appreciate the Phase I discussion. They were supportive during the talk and made good comments afterward. I felt like this was going to be a worthwhile trip.

The Visit — Phase II

The QAT was made up of 12 people. All were experts in area planning. The employee length of service ranged from three to 30 years. This was an energetic, extremely intelligent group. They appeared eager to get started, though somewhat apprehensive. This is usually the case.

The first step in Phase II is to ask the group to list the kind of problem areas they face in getting their job done. In about one-half hour, the group developed this list:

- Using obsolete methods.
- Insufficient teamwork.
- Group coordination.
- Changing product environment.
- Unclear direction/responsibilities.
- Lack of training.
- Lack of management support.
- Unclear job description.
- Unclear interfacing between organizations.
- Lack of communication and commitment.
- Lack of understanding customers' needs.
- Low or confused acceptance by the technical group.
- Multiple product lines' overcommitment.
- Control span.

In analyzing this list, you'll find that it contains more causes than anything else. The key question is, how are these causes manifested? What is the *effect* related to these causes? The answer in this case was that organization is not as effective as it could be.

Whenever you ask a white-collar group what kind of problems it is having, the group usually answers with the causes, as was the case here. When you ask what the result is the group defines the effect in the cause and effect analysis.

This is the key difference to using cause and effect other than in the classic way, when the effect is known. Ishikawa's well-known example of a known effect is the wobble of a machine. In the white-collar world, the effect is not as apparent. You must work with a group to define it. Once it is defined, it becomes quite apparent and understood by the group.

Some white-collar effects are:

- Late payments.
- Wrong terms and conditions.
- Unknown customers' needs.
- Disorganized work process.
- Inaccurate office records.
- Too much unbilled.
- Ineffective customer contact.
- Inadequate customer service.

The group does a cause and effect analysis once the effect is defined. The group defines each cause and uses the original brainstorming list as a guide. Sometimes all items are used; most times the original list is refined. When all the causes are defined, the group ranks them in order of importance for management. The rank order for this group is as follows:

(1) No clear direction and responsibility.
(2) Low or confused acceptance.
(3) Insufficient teamwork.
(4) Lack of leverage over supporting organizations.
(5) Lack of involvement in the commitment process.
(6) Lack of training.
(7) Confusion among levels of management.
(8) Too much reactive behavior.
(9) Insufficient resources.
(10) Different goals among organizations.

The self-correcting feature of the exercise came about during the force field analysis session when the group realized that numbers 7 and 10 were being addressed throughout the entire exercise and did not need to be handled separately. This analysis ended up with just eight defined causes.

The group developed eight force field analyses covering each of the causes listed. These are diagrammed in Figures 9 to 17.

Figure 9

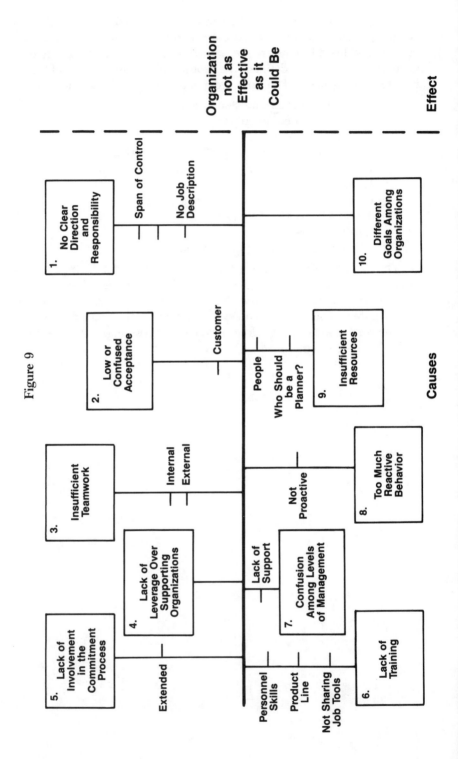

Figure 10

1 — Level of Clear Direction and Responsibility is too Low

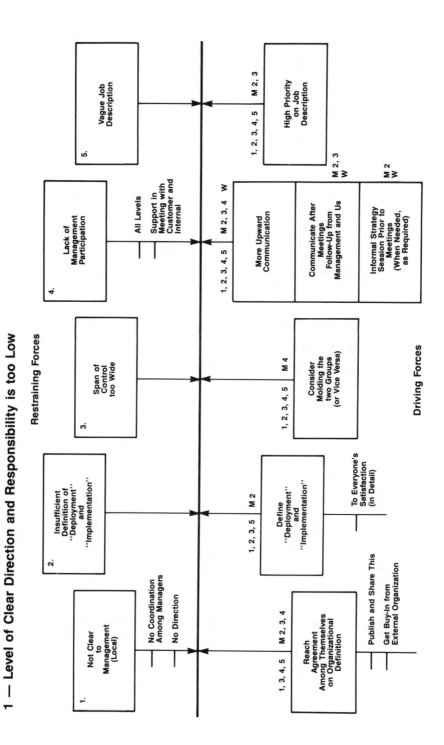

Figure 11

2 — Level of Acceptance and Understanding of Area Planning is Low

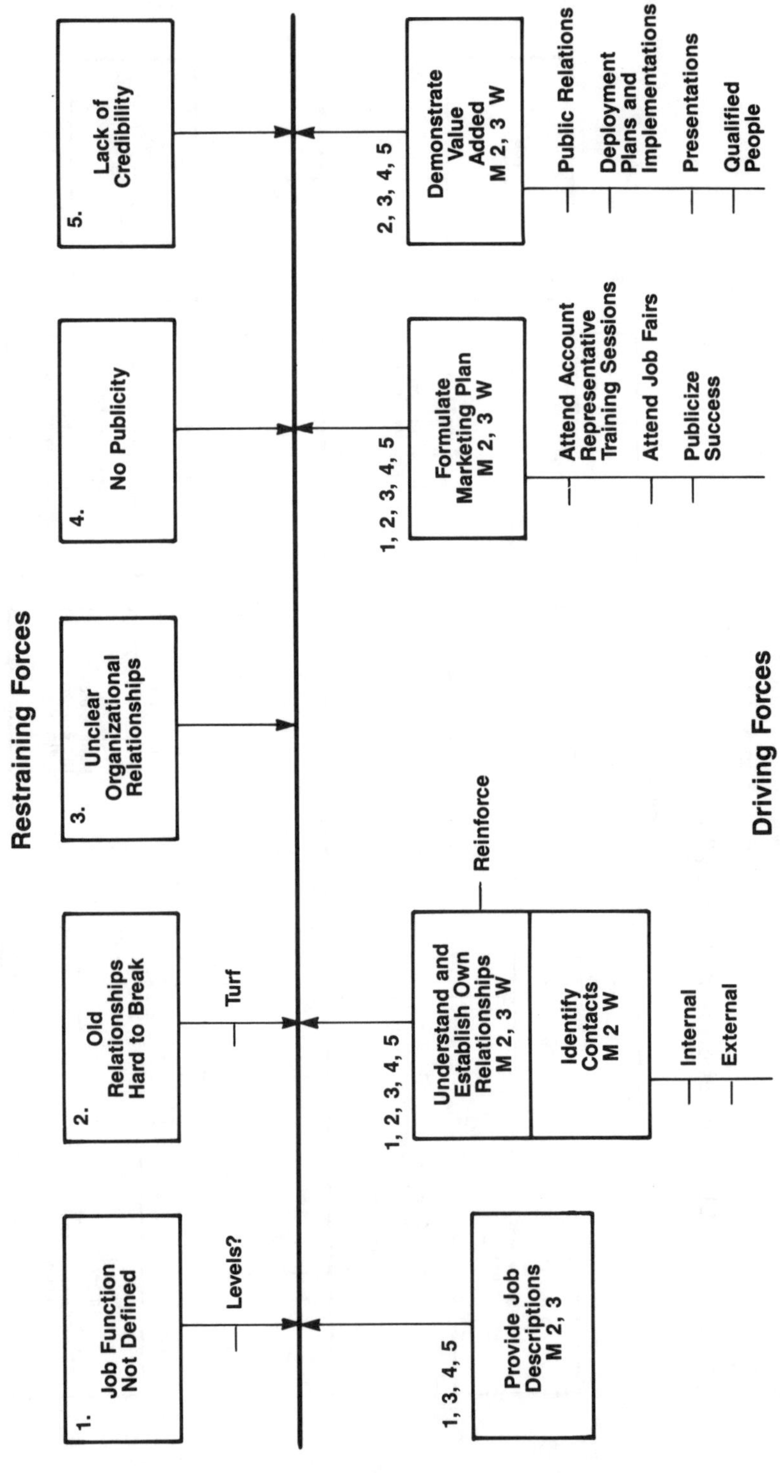

Restraining Forces

1. Job Function Not Defined
2. Old Relationships Hard to Break
3. Unclear Organizational Relationships
4. No Publicity
5. Lack of Credibility

Levels?
Turf

Driving Forces

Provide Job Descriptions M 2, 3 — 1, 3, 4, 5

Understand and Establish Own Relationships M 2, 3 W — 1, 2, 3, 4, 5
Reinforce
Identify Contacts M 2 W
Internal
External

Formulate Marketing Plan M 2, 3 W — 1, 2, 3, 4, 5
Attend Account Representative Training Sessions
Attend Job Fairs
Publicize Success

Demonstrate Value Added M 2, 3 W — 2, 3, 4, 5
Public Relations
Deployment Plans and Implementations
Presentations
Qualified People

Figure 12

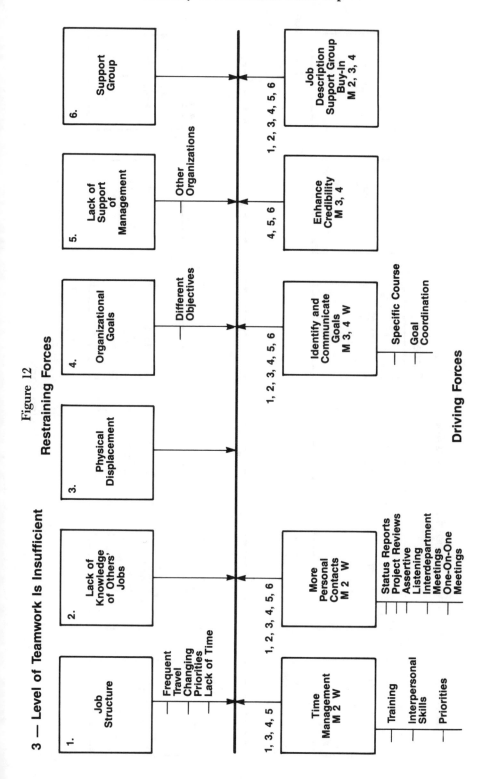

Figure 13

4 — Level of Leverage Over Supporting Organizations is Insufficient

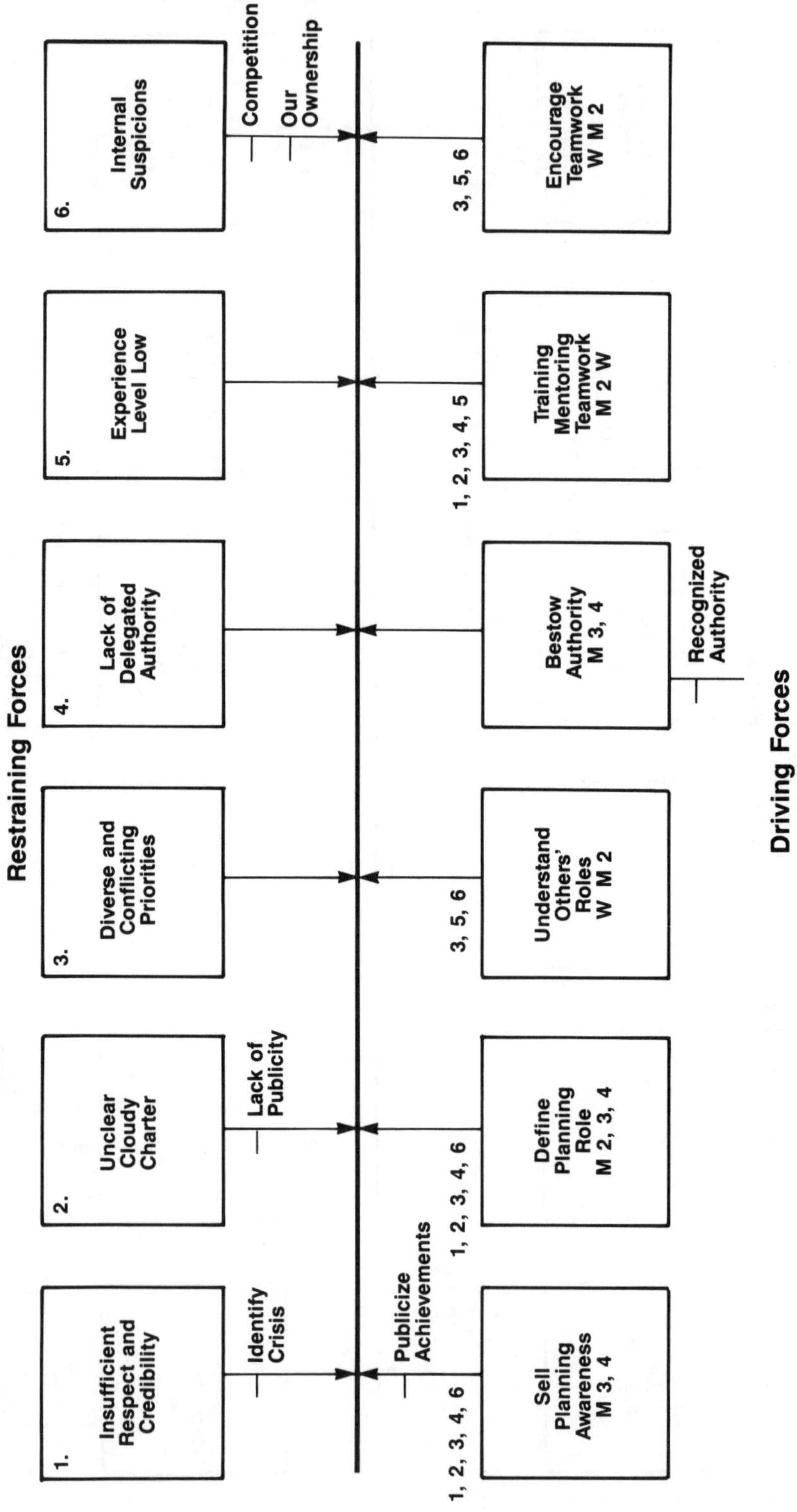

Restraining Forces

Driving Forces

| 1. Insufficient Respect and Credibility | 2. Unclear Cloudy Charter | 3. Diverse and Conflicting Priorities | 4. Lack of Delegated Authority | 5. Experience Level Low | 6. Internal Suspicions |

Identify Crisis

Lack of Publicity

Competition

Our Ownership

Publicize Achievements

Sell Planning Awareness M 3, 4 — 1, 2, 3, 4, 6

Define Planning Role M 2, 3, 4 — 1, 2, 3, 4, 6

Understand Others' Roles W M 2 — 3, 5, 6

Bestow Authority M 3, 4

Recognized Authority

Training Mentoring Teamwork M 2 W — 1, 2, 3, 4, 5

Encourage Teamwork W M 2 — 3, 5, 6

Figure 14

5 — Level of CPAC Involvement In Commitment Process Is Low

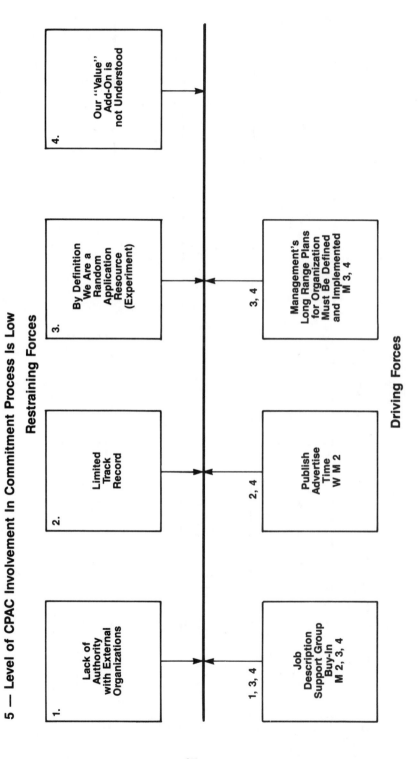

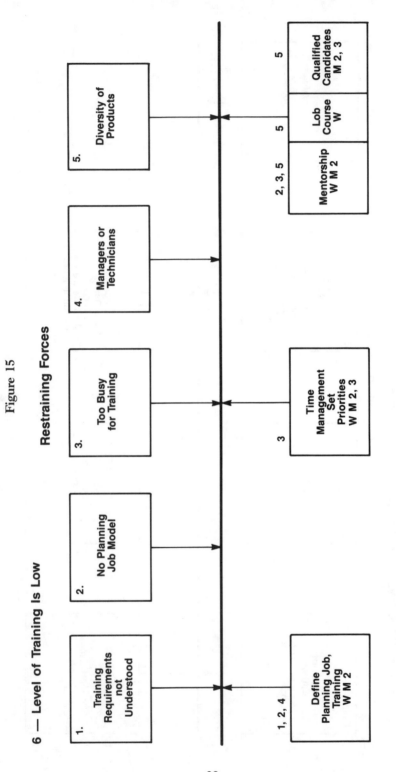

Figure 15

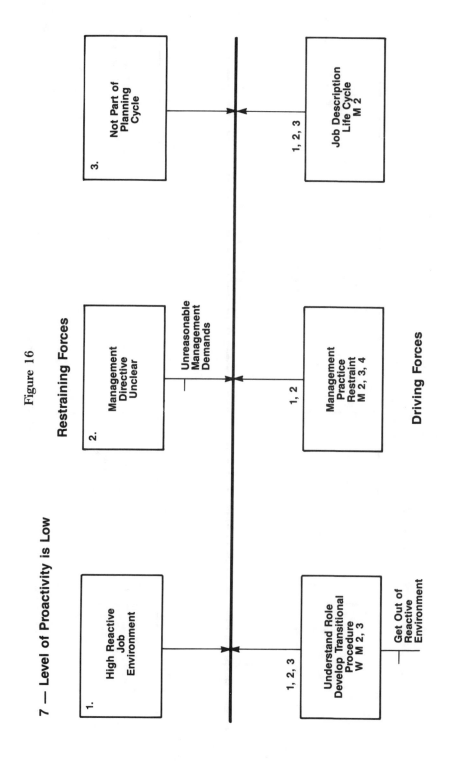

Figure 16

7 — Level of Proactivity is Low

Restraining Forces

Driving Forces

3. Not Part of Planning Cycle

Job Description Life Cycle M 2

1, 2, 3

2. Management Directive Unclear

Unreasonable Management Demands

Management Practice Restraint M 2, 3, 4

1, 2

1. High Reactive Job Environment

Understand Role Develop Transitional Procedure W M 2, 3

Get Out of Reactive Environment

1, 2, 3

Figure 17

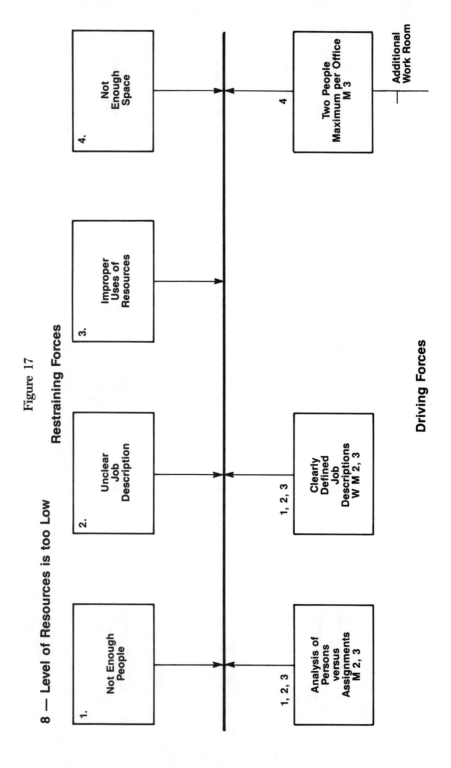

8 — Level of Resources is too Low

Restraining Forces

1. Not Enough People

2. Unclear Job Description

3. Improper Uses of Resources

4. Not Enough Space

Driving Forces

1, 2, 3 — Analysis of Persons versus Assignments M 2, 3

1, 2, 3 — Clearly Defined Job Descriptions W M 2, 3

4 — Two People Maximum per Office M 3

Additional Work Room

After all the force fields were completed, we summarized the driving forces into management (noted M) or worker (noted W). We asked the group to give its best idea of which management levels could help solve the problems (this is indicated by numbers after the letter M).

The numbers on the top left of each driving force are the restraining forces above that would be affected if the driving force was implemented. We then count the driving forces and find out how many are management, how many are worker, and which levels are involved. In this case, management was listed 37 times and the worker once. Worker was listed jointly with management 21 times. This meant that management, although listed in 99 percent of the cases, had the worker help in 37 percent of the cases. This was not surprising in a group where the workers were high level, highly paid, experienced people working on complex problems. We get much closer to Deming's 85/15 ratio in situations in production-type organizations.

You will recall, the facilitator actually did the cause and effect study and the first force field analysis then "passed the crayon" to the first person on the right or left side of the semicircle to do force field analysis number two. In this case, we have presenters for force fields two through eight. We do not have presenters for the cause and effect and first force fields. I always ask for volunteers and always get two. These are usually people who would have liked to participate, thought they were passed by, and then jump at the chance to be a part of the action.

So far, we have added up the management/worker tally and we have assigned a person to discuss the cause and effect and each force field analysis. We are now ready for the presentation to management.

The Visit — Phase III

As mentioned, this is an exciting event. Our group was relieved and ready; their humor was still very evident even though a week had passed since completion of Phase II.

I opened the session with discussions of the management versus worker responsibility summary; management responsibility was 63 percent, worker responsibility was 37 percent. Broken down further, management responsibility was 45 percent lower level, 35 percent middle level, and 20 percent higher level. This did not follow the normal 85/15 rule distribution because this was a relatively small group of technical experts with several years' experience. It's difficult enough to differentiate between worker and management in this highly professional world, so the 63/37 ratio is not surprising.

The next statement I make defines the purpose of this particular session. I told the group that this is an input session from the QAT to management. It's not the time for management to evaluate each statement or commit to specific action. This is the time to make sure management understands the input of the QAT and to clarify any questions.

I reviewed the original brainstorming list of problems and pointed out that all the items listed were causes. The effect of all the group-defined causes was that the organization was not being as effective as possible.

I introduced volunteer number one to walk the group through the cause and effect study, and to define each cause. The remaining force field analyses were covered by each QAT member who acted as facilitator during the QAT session.

After the eighth force field, I went to the easel and wrote the words *WHAT'S NEXT?* as suggested earlier.

I recommended that QRT be established, with a supervisor as chairman. The third-level manager volunteered, named his two immediate subordinates as members, and asked for four volunteers from the QAT and one representative from the staff group.

Eight people became members of the QRT; their first meeting was held immediately after the session. The true test of this entire endeavor will be what the QRT actually does in regard to the driving forces. It is too early to report results of this particular team, but we have shared results throughout this report that show continuous progress in process improvement.

PART FIVE
A Case Study
Introduction

A while back I used force field analysis to manage a complex corporate project. The mission of the project was to develop new ways to price our three major services. This may seem like a rather easy task at first glance. The complexities, however, of a supply contract that had been on the books for over 40 years, a highly authoritarian headquarters staff, and the three services being provided nation-wide by thousands of employees tended to make the task quite complex.

The first mission was to select people with expertise in each of the three services disciplines, as well as to obtain help from major corporate groups. A core group of 18 people was assigned. The group was assembled for a one-week orientation and goal setting and then dispersed in four major locations throughout the company. Three groups specializing in each of the primary services became known as *field teams* and the corporate group became known as the *consulting group*. The field teams embarked on a data-gathering phase that meticulously defined each element of the major services, analyzed the data to assure consistency of cost and price treatment, and made recommendations on how to improve consistency and make way for possible new ways of doing business with the ultimate customer.

The total group met periodically during the life of the project to ensure that each team was touching all the bases and that the overall mission was on track and making progress.

Very early in the life of the project, as project manager I felt a desperate need for help in managing the project effectively. Each field team began to take on its own personality, our corporate bosses were trying to unduly influence the outcome, specific team members began to develop weird traits and actions that were negatively affecting the project, and field locations management was beginning to become an obstacle to progress. I called a friend, Darrell Storholt, at our Corporate Education Center in Princeton and asked for help.

After describing all of the problems in more detail, Darrell suggested use of the force field analysis technique to get our hands on all the balls in the air. In about a week I had categorized each of the problem areas then in existence in terms of "level," discussed earlier in the book, and began to define specific restraining forces.

I didn't know about cause and effect at that time. Now I can see that each of the problem areas defined were actually "causes" and the "effect" was that the overall project was in jeopardy. The cause and effect study is shown in Figure 18.

The force field analysis that follows shows the restraining forces, the driving forces, and actual results of the actions taken. There is no question in my mind or in the minds of those in control at the time that the use of force field analysis actually saved the project. The recommendations coming from the project were presented to the highest level policy group in the entire company and were accepted. It is most interesting to note that those recommendations were so revolutionary that they form the foundation of ways of doing business today.

The point here is to explain the use of force field analysis, not to give definitions of the actual recommendations. Over time the recommendations may be augmented or changed completely; the use of force field analysis is the important thing for the reader to understand.

With this in mind, let's now examine each "cause" and its force field analysis.

Figure 18

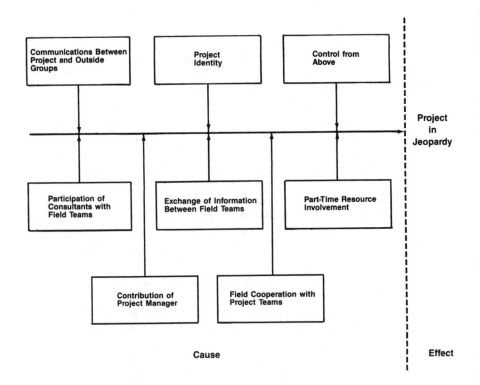

Examples of Force Field Analysis
Cause 1
Level of Control from Above

During the initial and early stages of the project there was continuous interaction between the project manager and those directing the project. Although the project got off to a good start after its members were told that control from upper management would be minimal, and that the project members could determine for themselves appropriate action to be taken, various actions led to deterioration of this position. Its overall effect on the project was detrimental and action had to be taken to alleviate the problem of control from above and can best be described in terms of the force field analysis made and the results that occurred.

Restraining Force. Pressure from above for control actually began before the initial start of the project in determining to whom the project manager should report. The approach was to open up the chain of command, make maximum use of whatever level would be necessary in solving the problem, and use the consulting team to solve project-oriented problems. These suggestions were accepted and used throughout the project. In essence, the general manager became the unofficial boss and provided guidance concerning administration of the project.

Other evidence of pressure for control from above can be summarized as follows:

- Pressure on individuals in the sponsoring organization to find out what was going on in the field.
- General concern about keeping closely informed on progress.
- Sponsoring organization attendance at the first project meeting.
- Insistence that the sponsoring organization be represented at the second project meeting in spite of strong feedback that attendance would be detrimental.
- During the project manager's absence early in the project the sponsoring organization enhanced the role of the consultants in terms of administration and control rather than technical content.

The above can be summarized as one restraining force indicated as pressure from above for precise control of the project from an administrative as well as content standpoint.

Driving Force. As mentioned previously, the driving force is a force applied to lessen the pressure and, therefore, aid in decreasing the level of control to a more acceptable level.

- A confrontation between the project manager and the sponsoring organization concerning the sponsoring organization's attendance at project meetings occurred over a period of approximately three weeks.
- A confrontation between the project manager and the general manager occurred; it was indicated that if the need for control from above con-

tinued to increase it would be necessary to replace people involved with the project. The general manager advised that this was not an appropriate solution, and that we should make the best of those presently involved with the project rather than expect to get top-notch replacements. The general manager indicated that there was no pressure on him to control or direct the project in any way.

- In line with the above clarification, another driving force was an announcement to the project manager that the need for a common control vehicle, common operation by all teams, and project identity at that stage had been lifted in favor of each team determining its own course of action in line with the objectives set at the initial training session. This announcement had a lasting and favorable impact on the total project in that it was a clear signal that the ball had been given back to the project to determine its own course. There was some apprehension about this. The group was given responsibility to decide the nature of future project meetings. Regular weekly progress reports were eliminated.
- A confrontation between the project manager and the general manager occurred concerning the role of consultants. It was agreed that the relationship of consultants to the chain of command should be "cooled off."

Results.

- The sponsoring organization agreed not to attend project meetings unless invited.
- Lifting the need for common control, common operation, and project identity had a positive effect on the operation of each team.
- Representatives from the three field teams met and decided on an outline for the final project report, assigned responsibility to individuals involved with the project for common sections of the report, and agreed on the makeup of these sections.
- Project identity evolved naturally among the three field teams. Agreement was reached on a common approach to identity functions in each of the three major areas being examined. Agreement was also reached concerning a tool to be used in defining the concepts, procedures, and implementation arrangements on practices and procedures.
- The general manager had a session with the consultants that helped bring the consultants' roles into perspective; i.e., representing their own discipline and helping and assisting the field teams at the field team's discretion. This was done in line with the project manager's recommendation noted above. In subsequent sessions between the project manager and the consultants, their role was further defined and clarified, and this was communicated to the three field teams. This resulted in a more positive contribution by the consultants to the project.

Restraining Force. The sponsoring organization attempted to organize its three representatives on each field team to meet periodically and discuss all aspects of the project and what was being accomplished by each team. This was also considered a restraining force keeping the level of control from above at a higher level than was acceptable.

Driving Force. The driving force to alleviate this was a confrontation between the project manager and the sponsoring organization that stressed the need for direct contact with each team concerning specific problems related to that team at the initiative of the sponsoring organization. It was indicated that the teams would take initiative when they considered it appropriate, but that it would also be appropriate for the sponsoring organization to make a contribution when, in fact, the sponsoring organization deemed it necessary. This approach was stressed both verbally and in writing to the sponsoring organization.

In addition, the project manager learned how the teams wished to interface with the sponsoring organization.

Results.
- The sponsoring organization agreed not to deal with its three team representatives as a group.
- It also agreed to deal with the individual teams on specific areas concerning contribution to content as a resource rather than on general philosophy or status.

Restraining Force. Another "group" within the sponsoring organization also expressed the desire to organize the same three team representatives to report to them occasionally on status and recommendations.

This was again leveled by a confrontation between the project manager and the group in which the suggestion was not supported, and another suggestion was made to cover this verbally with each individual team.

Results.
- The group agreed not to deal with the three team representatives as a group.
- Contribution by visits to teams as well as periodic discussions on a continuing basis were made by the group.

Restraining Force. The sponsoring organization's attendance at monthly meetings was discussed earlier under pressure from above for control.

Driving Force. The driving force was a confrontation between the project manager and the sponsoring organization.

Results. The sponsoring organization did not attend meetings.

Restraining Force. There was direct interaction with project members by upper management.

Driving Force. The project manager monitored assignments, supported assignments when they were appropriate, and changed them when they were not.

Results. The project remained on target throughout.

Restraining Force. Less guidance by the general manager contributed to the increase in need for control from other levels.

Driving Force. The project manager took the initiative to keep the general manager informed and sought advice where appropriate concerning project administration. Several key factors are significant.

The first was a suggestion to the general manager to meet with the training consultant and get an objective opinion concerning the progress and problems of the project. This meeting clarified the project manager's role in relation to the sponsoring organization. Another suggestion was given to the general manager concerning timing of his field visits and what he should look for in those visits. It was also suggested that feedback be given to the general manager concerning the disruptive influence of the consultants.

Results.

- Prior to the start of the project, concern expressed by the project manager led to securing more appropriate candidates to represent the sponsoring organization in the project. This set the pace for other organizations to follow and had a significant impact on the caliber of people selected to represent other disciplines as well.
- Clarification was provided concerning the suggestion to replace people. This course of action was not taken in favor of improving the individuals and letting them do the job themselves.
- The session with the training consultant clarified the role of the sponsoring organization to be resource-oriented and the role of the project manager to be administration-oriented. The general manager evolved here as the administrative as well as the functional head of the project and the project manager's boss.
- The general manager agreed to replace one team member.
- The general manager made sure the project manager was well informed concerning his field visits both before and after the visit. The project manager initiated contact in *all* instances.
- The general manager took appropriate and timely action concerning definition of the consultant's role.

Cause 2
Level of Project Identity

Restraining Force. In the early stages of the project, the field teams were mainly concerned with geographic moves: getting situated at the new location both at home and at work, determining a plan of action for their individual team in the data gathering phase, and then actually going out in the field to interview key people and organizations in accomplishing the data-gathering operation. These kinds of activities naturally required close coordination among the four people on each field team and did not, in fact, require any major degree of coordination or cooperation between teams. Although a natural condition, this fact was thought of as a restraining force that tended to decrease the level of project identity to a point below that which would be necessary in the development and analysis of actual alternatives that would occur later in the project.

Evidence of high team identity includes the following:

- No meaningful communication between teams in the early months of the project.
- Concern of some teams about what the other teams were actually doing.
- No meaningful use of consultants by the field teams until later in the project.
- A complete takeover by one team in an early project meeting with little or no evidence of discussion regarding participation of all teams before the meeting.
- A tendency of one team to ignore the planning done during the training week concerning schedules and target dates.

Driving Force. An initial reaction was that strong team identity would be detrimental to the project in its early stages as well as in later stages. This was emphasized at the first project meeting. After lengthy consideration and discussion with various management levels as well as the management training organization, it was concluded that high team identity was a natural process in the early stages due to the kinds of activities indicated above and that any force applied by the project manager to unify actions of the teams and make them compatible would be inappropriate. This posture was announced at a project meeting where the need for common identity and control vehicles was lifted. This recognition can be considered a driving force. It was considered that the pressure of time in development of the final package would require total cooperation and the melding of the teams at a later stage in the project.

Results. The project manager and training consultant visited two field teams and discussed communication problems between teams at the initiative of both teams. As a result, five members of each field team initiated and attended a special meeting representing the three field teams and agreed on the final presentation format. They also assigned specific individuals to be responsible for the common areas of the report, and assigned teams the responsibility for sections of the report involving individual team recommendations, implementation arrangement, and a summary section. The group agreed to specific due dates for the outline, first draft, final draft, and final document dates.

Although the meeting only concentrated on common areas and did not delve into the specifics of actual recommendations, it was the first meaningful attempt by participants across the project to get together on their own and work toward a compatible end result.

The second major result that led to an increase in the level of project identity was the evidence of sincere sharing that took place midway into the project. Particular emphasis was devoted to each team's approach in arriving at alternatives and conclusions. Part of this discussion was the development of a matrix including consideration of all elements as well as implementation and project evaluation of alternatives. It was indicated by the group that maximum communication would be vital throughout the remainder of the project, and that all teams should continue to strive for compatibility and a continued common approach. The field teams also gave indication that they were dependent on the consulting team for help in various key areas.

A major concern of the project manager at this point in the project was the possible tentativeness of the outward support for maximum communication between teams, and a desire toward development of a compatible approach.

Restraining Force. It is a main concern of the project manager to influence the teams to get together. See the discussion starting on page 59 for a complete description of the driving forces and results.

Restraining Force. Activity of teams in the early stages of the project requires and enhances team identity. The restraining forces mentioned are appropriate.

Driving Force. The driving force was to respect this as vital in the first half of the project. It was decided that project identity would be more important in the later stages.

Results. Project identity grew steadily throughout the project.

Restraining Force. The teams were reluctant to adopt similar methods of attack due to the individuality of team experiences in the data-gathering stage of the project and, therefore, the lack of real need for communication between teams. Different methods of attack were established to do the actual data gathering.

Field Team A began almost immediately after a brief period of determining what kinds of questions would be asked, the purpose of the meeting with field representatives, and who should make visits and who should be interviewed. As a result Team A placed major emphasis on revising its questions and approach to field visits after initial field visits. It remained on target concerning schedules of visits in line with their original planning. However, in retrospect, Team A concluded that fewer visits would have been sufficient to do the job and so advised Field Team B concerning its planned visits to the field.

Field Team C spent more time developing an extensive questionnaire that it, in turn, sent to key people. This questionnaire received abundant attention and criticisms from many different sources. Because of this, the questionnaire was withdrawn and Field Team C interviewed key people in an effort to answer pertinent questions and concerns. This attempt was not totally satisfactory because of the approach used by Field Team C in obtaining information from those interviewed. Negative feedback was provided to Team C by the project manager concerning this development as it occurred and as the team was interviewing in the field.

During the first four months, Field Team B decided to place maximum emphasis on developing a firm base of understanding concerning detailed practices and procedures, and its effect on all related organizations. This also included determining the impact of a group studying and developing national guidelines and also a group working on the same endeavor from a divisional standpoint. They tested their ideas on several selected locations before actually developing a comprehensive questionnaire.

Field Team B decided to visit all field locations (45 visits). Much pressure was brought to bear on Team B by the project manager to decrease the visits based on Team A's advice and experience. Subsequently, Team B decided to decrease visits to 25 rather than 45 locations.

Feedback from several key field locations, including an on-site visit, during one interview indicated that the planning time spent by Field Team B was valuable in terms of quantity and quality of information received.

Driving Force. The fact that the teams went in different directions during the data gathering stage and were reluctant to adopt similar approaches was not thought of as a major deterrent to getting the data-gathering job done. This, of course, would change as the effort of each team changed in the actual analysis and development of alternative stages. This was suggested informally to each team at various times and was also mentioned in terms of the kind of meeting that would be held at midpoint. In spite of strong feeling on the part of some project members that results could only be achieved at a project meeting by breaking the group into smaller groups, the project manager believed that the major results would have to occur with the 18 project members working together to arrive at conclusions as to the approach that would be adopted by the project as a whole in the development of actual alternatives. This feeling was supported by a few key project members and actually did occur.

Results. All teams adopted similar methods of attack. Field Teams A and C met two weeks before the project meeting and were well along the way to adopting similar approaches and assuring compatibility.

Field Team B came to the project meeting looking for help as a result of devastating feedback to the team from the project manager regarding the general manager's visit. They received help by way of constructive criticism from each field team.

Restraining Force. There was a feeling of competition between teams. From the start of the project there was an inherent feeling among all three field teams that the end result of the project would be measured in terms of what each team provided rather than what the total project provided as an end product. This led to the lack of communication between teams in the beginning of the project and the desire to come up with the best plan to get the job done first.

Driving Force. This was first brought up early in the project but clear action was not taken by the teams at that point. Agreements were reached to exchange all written material, use conference calls more frequently between teams, and communicate more in setting up meetings. It was also recognized that working closer together would be important to achieve meaningful end results.

The project manager began to identify common problems that required communication between teams as they developed. The key item here was the need to develop a common outline.

Results. Teams interchanged written material and started to discuss common problems more frequently. They achieved agreement on the common outline and the framework for the approach to development of actual alternatives. This was tested from a technical standpoint by a visit made to each field team by the sponsoring organization management. Strong suggestions were made for closer collaboration between teams concerning actual alternatives. This actually occurred and continued throughout the life of the project on a very healthy level.

Cause 3

Level of Communication Between the Project and Outside Groups such as Consultants and Professors

At the outset of the project it was indicated that it would be feasible and, in fact, compulsory to get advice and consultation from organizations and people outside of the company. This would include consultants from both the legal and academic fields. In addition, it was suggested that the project could benefit from finding out how the job is handled in other industries and this could be determined through actual interviews with these companies. The project members believed this to be a good idea and placed much weight on it in the determination of final alternatives. This was evident in the planning done during training week where much discussion ensued concerning whether this was a field team responsibility, a consulting team responsibility, or a joint responsibility. The level of communication between the project members and people outside the group was restricted in the first month of the project as follows:

- No interview of outside companies.
- No outside consultants.
- No public relations.

The following will describe the restraining forces and driving forces that have eliminated any communication between barriers between the project and outside sources.

Restraining Force. Legal restrictions.

Driving Force. Confrontation with the director to obtain modification of restrictions.

Results. The director modified all restrictions previously. The modifications, however, did not produce any actual communication between the project and outside groups.

Restraining Force. Teams disbelieved the modifications. Although each category was modified to some degree, project members did not pursue any actual knowledge from outside sources at the onset.

Driving Force. Teams gave specific definitions of their needs. On several occasions the project manager tried to get the teams to take advantage of the modifications and make specific recommendations concerning their needs. This was done in the early stage of the project by each team concerning subjects to be discussed with an outside expert. Field teams requested that a consultant team member seek information through the company about the procedures of other organizations.

Results. As indicated, a consultant team member was assigned to obtain information concerning other organizations. This was pursued and results were available to each team.

Restraining Force. There was a high level of control by the sponsoring organization. Due to the sensitive nature of the information being sought, it was decided that the director would be a control point concerning such interchange and data. This is thought of as a restraining force, although a necessary action.

Driving Force. The control at this level was respected as vital and teams were encouraged to report requirements specifically and directly to that level through the consultants. A serious attempt was made here to make sure that this interface dealt with the subject of communication to outside organizations and did not get involved with administration of the work of the project per se.

Results.

- The consultants dealt with the director regarding the use of outside consultants.
- The teams worked through the consultants.
- The team members gained an appropriate level of understanding concerning the need for this control.

Cause 4

Level of Participation of Consultants with Field Teams

This subject received much discussion during training week. At that time, there was a strong feeling among the three field teams that the consultants were trying to put themselves in a supervisory or controlling position concerning review and evaluation of the field teams' work. This was not completely reviewed during training week and remained a consideration to be dealt with throughout the life of the project. It is important at this stage to appreciate the role that was defined by the consulting team during training week and agreed on by the entire group.

Restraining Force. The team believed that the consultants must review and pass judgment on team activities. As indicated above, this began during training week and was prevalent in some degree throughout the remainder of the project. However, the consultants continuously felt that they had a responsibility to oversee the overall project activity to make sure that end results would be practical. This is a delicate task; actions and words could lead field teams to believe the consulting team's role was more controlling in nature than it actually was. As indicated previously, the consultants went overboard at one point which supported the feeling of the three field teams. This feeling and the inappropriate action of consultants was considered a restraining force on the actual contribution of the consulting team.

Driving Force. Teams should be convinced of the real role of the consulting team through major contact with field teams during the development of alternatives stage. This contact essentially came from two major sources: strong pressure from the project manager on field teams to make maximum use of the consultants during the development of alternatives stage, and maximum penetration by the consultants with each of the field teams during the same stage.

Results. Throughout this section field teams have been referred to as all having the same problems. In actuality, there were great differences in attitude among field teams in that some used consultants to maximum advantage sooner than others. Because of the teams' early suspicions that the consultants were trying to take over and the fact that these suspicions were confirmed, the consulting team's effectiveness was seriously hampered. One team, for instance, virtually ignored the potential contribution of two consultants. In the earlier stages of the project, the field teams provided the consultants with specific tasks. This helped to define the role of the consultant specifically.

Restraining Force. There was a lack of specific tasks from field teams to consultants. This was a problem in the earlier stages in the project.

Driving Force. More input from field teams to consultants defined specific tasks.

Results. This input was provided and appropriate action was taken by the consultants.

Restraining Force. There was less direct contact by consultants with field teams during various stages of the project. This was more apparent in the earlier stages of the project when the field teams were interviewing and collecting data, than it was in the latter stages of the project when the actual development of alternatives took place and major activity was started to test for practicality and implementation capability by the consultants themselves.

Driving Force. Encourage more field visits by consultants were encouraged, and field teams were encouraged to work with consultants when appropriate. Consultants were assigned specific responsibility for preparation of the general manager's speech. Three consultants were assigned to individual teams.

Results. This was more of a problem with Field Team A than with Teams B or C. This was primarily caused by the fact that Team A prescribed a task to be done by the consultants from early in the project that was not deemed a necessary task by the consultants.

They accepted responsibility to do the job, negotiated for technical assistance from the Corporate Statistical Analysis Organization, and provided Field Team A with substantial data that were used as a base by Team A in answering the original question. This conflict began to subside when Team A realized it would need support, help, and advice from the consulting team.

Restraining Force. The earlier stages of the project were exclusively involved with gathering data. The consultants' main contribution to field teams in the earlier stages was to help them gather data in areas that the field teams believed could be done more appropriately by the consultants.

Driving Force. The project manager decided to wait until the development of alternatives stage of the project before suggesting action on the part of the consultants that would contribute toward successful completion of the project. It was thought here that the field teams would come to realize that the contribution of the consulting team was important to their individual progress. The driving force was allowing the pressure of time to take its course rather than forcing a false and, therefore, insignificant interface between the field teams and the consulting team.

Results. A great change in attitude among the three field teams started midway during the project and continued.

The consultants contributed significantly in the development of alternatives and their crystallization in the final proposal.

The consultants activity increased significantly through the life of the project in the area of testing for practicality and actual implementation of final alternatives.

Restraining Force. Field team identity was strong in the early stages of the Project, and, therefore, the need to identify with the consultants was weak.

Driving Force and Results. See previous comments on this.

Cause 5

Level of Exchange of Information Between Field Teams

Because the individual field teams thought they had enough autonomy to come up with a final product without interface with other teams and without interface with the consulting team, there was less communication than was actually required during the earlier phases of the project between all teams. As indicated previously, this was natural during the data-gathering stage in that various approaches could, in fact, be used to accomplish this particular phase.

Very early in the development of alternatives phase, the lack of adequate communication was recognized by teams as being a serious detriment to the accomplishment of final objectives. The following forces describe attitudes and resulting actions that, in fact, contributed to increase effective communication between all teams.

Restraining Force. There was a feeling of competition between teams. This has been described adequately in previous sections.

Driving Force. The field teams must realize that the feeling of competition exists and that all negative aspects should be eliminated by the teams themselves. This posture began to take place when the problem of competition between teams came up in discussion. At that time, it was agreed that more information should be exchanged, both verbally and in writing.

As the alternatives for Field Teams A and C began to take shape, the project manager stressed compatibility of approaches in his discussion with each team.

Results.
- The competitive attitude lessened to a considerable degree and was most noticeable at key points during the life of the project.
- In addition, the field teams started to visit each other to work toward a more compatible approach.

Restraining Force. Teams were too busy solving their own individual problems in the early stages of the project.

Driving Force. It was thought that the team would recognize the need for a joint thrust on specific areas later in the project.

Results. The elementization and partitioning Field Team C originated, the manner of illustrating alternatives in matrix form as developed by Field Team C, and Field Team B's approach of determining advantages and disadvantages for establishing their alternatives are best examples of this.

Restraining Force. There was a basic feeling of difference in competence among teams. In the early visits, the project manager got the impression from various individuals that a feeling existed concerning the difference in level of competence of individuals on other teams. Specifically the degree of education was mentioned. One team had four members one of whom had a bachelor's degree, one a master's degree, and two high school degrees as compared to another team whose four members each had master's degrees, including one who had a law degree. Other concerns included the success or failure of teams in their interviewing of field locations.

Driving Force. In each case, the project manager found that one team's evaluation of another in any of the areas mentioned was usually overplayed for reasons that were not quite clear. These reasons could have been the restraining force of unhealthy aspects of competition between teams. The concern about the difference in education, for instance, was definitely the case of pitting one team against another instead of figuring out a way to make maximum use of the educational level on other teams to the benefit of those that, in fact, required it. The project manager's approach was to indicate to each individual that had this concern that the gap between teams, whatever it may be, was not as severe as it may have seemed. He encouraged maximum contact both informally on a one-to-one or team-to-team basis, and also formally at project meetings. The teams were also encouraged to participate more fully in the preparation of agenda items for each meeting.

Results. All talents across the project were used to maximum efficiency in coming up with the end report. This affected the field representatives on each team as well as those representing common disciplines. In the last five months of the project, the consultants were effective in having positive impact on the teams, the general manager's speeches, the final report, and their own home organizations concerning implementation effort.

A healthy type of confrontation also developed between teams concerning problems.

Restraining Force. Teams did not share basic data and ideas in early stages of the project. This was covered in previous comments.

Driving Force. Teams need to recognize this and take action to solve it themselves. An example can be set by sharing information both verbally and in writing.

Results. See previous comments.

Cause 6

Level of Part-Time Resource Involvement

The part-time resource people consisted of representatives from eight major organizations in the company. The training consultant was involved more fully than any other part-time consultant throughout the life of the project. The other part-time resource people were active in varying degrees throughout the project. The level of involvement, however, was a concern in the early stages for reasons that will be evident in the discussion that follows.

Restraining Force. The field team wanted to be autonomous in the early stages of the project. An initial observation of field teams was that they could pretty much handle the task without too much help from either full-time or part-time resources. There was a strong desire to do the job themselves instead of relying on help from other organizations.

Driving Force. The field teams needed to be told that the part-time consultants could do more to support their efforts. The consultants would be a good source for ideas and also would help prepare the way for possible implementation. It was also indicated to the teams that the consultants were really not contributing as much as they could.

Results. All part-time resource organizations became actively involved throughout the life of the project.

Restraining Force. The part-time consultants were somewhat reluctant to initiate help.

Driving Force. A meeting was held with all consultants. Two-way communication was stressed at this time. The general manager participated by encouraging the consultants to initiate help on their own.

Results. There were no apparent problems of lack of support or communication between the part-time resource people and the project. All communication that had been made was positive and help had been obtained wherever required.

Restraining Force. The training consultant was a major participant in the December Training Session held at the Corporate Education Center. However, his efforts in coordinating with the project were minimal early in the project. This changed considerably when pressures were recognized by the project manager.

Driving Force. The project manager had a session with the training consultant indicating need for more contact with the teams.

The project manager kept the training consultant informed with all written material concerning the project and by phone conversations seeking advice or advising on key points throughout the project.

Results. The training consultant attended a key project meeting and was a help in advising key action concerning performance of specific individuals, including the newest member of the project. He advised the project, at that time, that he would be available for consultation on an individual or team basis if desired. He visited the three field teams and offered to discuss problems or concerns about communication between individuals on each team, communication between teams, and consideration of individual growth. This led to a conclusion by both Field Teams B and C that a special project meeting would be appropriate to iron out problems concerning the final project report, what it would include, and who would be assigned responsibility for specific sections.

The training consultant had an extensive private session with the general manager that developed into a discussion participated in by the four levels of management in the sponsoring organization. The result of these combined sessions was that all levels gained a clear understanding of the role of each person and the contribution to be made to the project. Essentially, it was agreed that the project manager would have full administrative responsibility for the project including schedules, objectives, personnel, approach, training, selection, and control of the project, and that the sponsoring organization would have responsibility concerning contribution to and evaluation of the specific content that would be developed by each field team. The fact that the development of content was not scheduled to start until the latter part of the project was stressed, and also that contact with the sponsoring organization during the data-gathering stage would be minimal except for acquiring data within that organization, which had already been accomplished.

The training consultant advised individuals and teams at various points in the project. This tended, in some cases, to improve individual performance, and in other cases, team performance.

As mentioned earlier, the training consultant suggested the force field analysis technique in analyzing various project manager concerns, which helped determine the cause of certain problems and actions that could be taken to alleviate them. This contribution was considered vital to the overall success of the project.

Cause 7
Level of Contribution of Project Manager to Teams

The matrix management combined with a synergistic approach required the project manager to develop a more liberal management style. The Theory Y approach was also appropriate due to the nature of the task. A word about each of these management techniques is appropriate.

Matrix management required that a chain-of-command approach not be used within the project itself. In other words, there was not one individual on each team that reported directly to the project manager concerning results of that particular team. The aspect we are discussing here is the fact that the project members worked in an atmosphere where rank or structure of any kind was not recognized. It was, therefore, important that the impact of the project manager on the various teams be concentrated more toward the developmental mode of operation than either strict control or the opposite which would be relinquishing responsibility for reaching objectives. The developmental mode placed major emphasis on informing, persuading, exploring, coordinating, stimulating, and sharing problems and solutions. Little emphasis was placed on enforcing, dominating, selling, or fighting to maintain maximum control.

The *synergistic approach* mentioned previously makes achieving results through group effort the maximum goal. The definition of a synergistic approach is "the cooperative effort in which the effects are greater than the sum of each part (individual working alone)." This, in effect, is the major reason for lack of structure within the project itself, and securing people with the best talents to represent specific disciplines that could, in fact, meld together as a team as well as a total group of 18 in adjusting themselves to the problem at hand and, in fact, solving it.

The Theory Y approach mentioned earlier refers, of course, to McGregor's more enlightened view of the employee by management. This theory assumes that people generally seek responsibility; that they have a positive capacity for exercising imagination, ingenuity, and creativity; and that they can exercise self-control. Although McGregor's theories address themselves specifically to the average person, the people selected for the project had high credentials in education and/or experience as well as, for the most part, a strong desire to participate in the project.

The following will describe forces that developed that restricted the project manager's role in regard to the above areas, and those forces that were applied to counterbalance them and allow a more adequate contribution.

Restraining Force. Direction from above equates to more control by the project manager in terms of enforcing and dominating. Cause 1 covered most areas where this was predominant; namely, the pressure for actual control of activity within the project from above, the desire of two separate organizations to organize the three people representing the sponsoring organization on the project, attendance of sponsoring organization at meetings, direct interaction with project members, etc. These actions tended to increase the control aspects and, therefore, decrease the developmental mode of operation that was thought of to be more in line with the matrix concept, a synergistic approach, and the Theory Y concept.

Driving Force. It was necessary to separate the functions of direction from participation in the content concerning activity from above. As indicated in Causes 1 through 4, attempts were made to involve teams with participation concerning content and any help that the chain of command could give in this area. Confrontation was the course of action taken when any kind of control or action was considered inappropriate.

As indicated above, the sponsoring organization's involvement was limited in the earlier stages when not appropriate, and sought in the later stages when it was appropriate.

Results. Causes 1 through 4 indicated results that were achieved as a result of the preceding driving force. In addition, the project as a whole was totally responsible for establishing a plan of attack including objectives for the entire project and gathering data from various field locations. The sponsoring organization did get involved in the development of alternatives stage and continued throughout the life of the project.

Restraining Force. More overt action by the project manager to control specific activities of teams led to less contribution and support of field teams in the early stages of the project.

Driving Force. A decision was made by the project manager to give maximum freedom to the field teams early in the life of the project. It was decided that the quest for a project identity, a common control vehicle, and adoption of similar approaches would be dropped — this resulted in the field teams developing their own system for controlling these activities. In addition, the need for a weekly progress report was lifted based on field team requests.

The project manager provided direction to individuals and teams as well as the chain of command when appropriate.

Results. As indicated, lifting the need for the items mentioned previously resulted in solid response by the field teams.

The following results highlight some major areas where direction by the project manager was effective:

- Field Team B reduced the number of its visits from 45 to 25; this facilitated meeting project schedules.
- One team member was replaced due to health. The replacement selection facilitated meeting project schedules and also aided the approach, content, and ultimate testing of alternatives for both Field Teams A and C.
- The decision not to replace others facilitated the end product and also helped the involved individuals.
- The need for compatibility was stressed to both Field Teams A and C. This resulted in a strong cohesive plan of attack by both teams.
- Team B was advised concerning the appropriate approach to solve the problem of developing only one alternative. This resulted in the development of several approaches.
- Consultants were advised concerning their role in the development of the interim status report. This resulted in a favorable reaction by field teams and an end product that was highly acceptable.

- Information was provided to the sponsoring organization, resulting in the following action:
  - The need for compatibility between Field Teams A and C was stressed. This was indicated in visits by the general manager and had a significant impact which also resulted in compatibility being established as noted previously.
  - It was indicated that Field Team B was recommending one approach and that decisions should be made concerning this approach. This also resulted in the several approaches being developed by Team B as noted previously.
  - General manager and director visits were recommended and arranged during the early stages of the project through the development of alternative stages. These visits had a significant impact on the quality of work and the content of alternatives. At a later stage it was recommended that the director be more specific concerning the appropriateness of actual alternatives.
  - A system of rewards was developed stressing informal involvement of the general manager in regards to the short-term goals of each individual. This was met with good feelings on the part of project members in that they realized someone on a significant level was concerned about their individual career goals and was willing to be a "willing negotiator" if this could, in fact, help them along the way.
  - As a result of a request for more meaningful involvement of the sponsoring organization concerning content, the director visited each field team and had two of his organizations explore the appropriateness of alternatives from their own organization's viewpoint.
  - All the results indicated previously evolved from the direct or indirect action attempted throughout the life of the project by the project manager.

Restraining Force. A strict chain-of-command approach lessens exposure of the teams to needed input. The best example of this is access to higher levels of management for required input regarding past agreements, current activity, and interpretations of policy and procedures.

There seemed to be a reluctance of individuals and organizations to initiate contributions of the aforementioned nature to the teams. As mentioned previously, the main problems here seemed to be the continuous confusion between direction of what the teams should do and input in the form of suggestions, advice, and criticism concerning content only. The project position from the beginning was to deemphasize the former in favor of a strictly project oriented direction for the project and to favor maximum interface with all organizations involving content.

Driving Force. It was necessary to cut through the traditional chain-of-command approach and go directly to the person or persons necessary to get the job done. This related to all organizations indicated previously as part-time resources, as well as those considered full-time resources.

Conversations were used, for the most part, to initiate the above kinds of action although correspondence was initiated at times to serve the same purpose. Both were effective and produced the necessary results.

Results.

- Visit to Field Team A shifted emphasis from detailed problems and concerns to overall problems and concerns.
- Visit to two other organizations stressed need for compatibility between both teams concerning approach.
- Visit to Field Team B indicated the inadequacy of a one-alternative approach in favor of a multiple-alternative approach.
- Visit to each team cleared up the problem of understanding the actual alternatives suggested concerning reliability and practicality from a sponsoring organization standpoint.
- Progressive interface between teams and all the full-time and part-time organizations resulted in better communication and understanding of the alternatives.
- There was a savings in time and money on the part of the project by making maximum use of even intuitive direction received as a result of all visits by higher management. The project manager in these cases utilized the technique of maximum communication between those who visited and those who were visited. This involved clarifying interpretations from both sides as to what was discussed, why, and what action should be taken.

Restraining Force. Pressure was exerted on teams and individuals at times to act in specific areas that may or may not have been appropriate. Examples of this include:

- The consultants were requested to visit Field Team B and determine the appropriateness of interchange of data and further interface between another organization that was allegedly studying the project problem and Field Team B.
- A consultant was asked to perform a task that the project manager was given but had not completed.
- Organizations endeavored to organize the three project members from the sponsoring organization on each team for purposes of reporting to them concerning progress and results.
- The consultants expanded their role to administrative actions.
- The introduction of a new manager led to some confusion regarding the role of the management team. This was temporary in nature and lasted only two weeks.

Driving Force. In all situations, there were attempts to accomplish results and relieve pressure on the teams at the same time. In many cases this clarified what was actually required and specific action or needs were communicated.

Results. The consultants decided that Field Team B had secured necessary input to be knowledgeable about the other organization's activity and also found that the team had elected to maintain an arm's length relationship in that this group had committed itself to a specific approach which in the end could only be considered among many approaches and evaluated at a later date in that framework. The consultants usually discussed jobs that were given to them and made sure that two persons were, in fact, not doing the same thing. Interface with the sponsoring organization was accomplished through other means than organizing the three project members on each team to periodically report in. The consulting role that was heavily leaning toward administration was clarified and changed. To firm this up, discussions were held with the consultants and also assignments were suggested concerning specific interface with each team on the special preparation work.
The role of the new manager was clarified and remained consistent with the role established previously.

Restraining Force. Playing a directive role at meetings was considered less than appropriate in achieving meaningful results. It is a natural tendency for the chairperson of a meeting to enforce or dominate when the going gets rough.

Driving Force. Problem sharing and stimulating maximum discussion was stressed whenever possible.

Results. Some meetings were more productive than others, but the overall trends seemed to be the ability to identify, share, and solve problems in a more efficient manner as time went on. *There was no dramatic difference, however, between the kinds of problems and confrontations that were experienced in reaching solutions during the first week of training and those encountered during the life of the project.*
The individuality of each person remained a very strong force opposing complete unanimity on any given subject. This did not deter the group as a whole in defining problems, arriving at solutions, defining objectives, planning detailed work to accomplish objectives, and producing a significant end result.

Restraining Force. There was continued misunderstanding of the consultants' role by the field teams as well as the consultants themselves. This problem continued throughout the entire project. The main problem here was the inherent fact that the consultants not only represented their own specialty, but necessarily got involved with overall problems from time to time that were of global nature to the project. This was necessary because there was either no one in the project to represent a particular discipline that may have been required, or the problem was general in scope and required a technical type response and/or solution that the consultants could offer in a natural way.

Some of the problem related to misunderstanding is mentioned above, but there was also the problem of competition and strong team identity that also tended to lessen the role of the consultants rather than enhance it. This was mentioned in detail in Cause 4.

The consultants, either through misunderstanding of their role or other pressures, sometimes made premature plans to leave the project or became involved as indicated earlier in an administrative capacity.

Driving Force. The major driving forces to counteract the above restraining type forces were to clarify the consultants' role continuously with the individual field teams when necessary, suggest action to the consultants individually or as a team when appropriate, support appropriate actions by either the field teams or the consultants concerning this specific problem, or attempt to provide constructive criticism when necessary.

Results. Consultants were persuaded by field teams that their presence and action was vital to the project in the training week and later during the project. During the training week, two consultants believed that their efforts were not required and volunteered to abort the project. This never did happen as a result of the interaction between the field teams and the consultants noted previously. An identical feeling occurred later from the same two consultants. Action was taken to clarify the role again including conversations with the project members, the general manager, and the home organizations of the consultants involved. This resulted in holding the consulting team together for the remainder of the project and benefiting from the contribution of the individuals involved.

As mentioned previously, the consulting role had to be defined and clarified. This was accomplished and communicated to each field team, and resulted in a more meaningful stance and contribution by the consultants.

See Cause 4 for more detail concerning the role of the consultants in the project.

Restraining Force. Another restraining force was the inability of a minority of members to direct themselves or handle the inherent conflicts of matrix management. This had a tendency to impede team progress at times when major attention was placed on individuals' problems and maximum attention to these problems rather than project objectives. Included in this was some confusion when team members did not have a technical boss to go to to solve a specific technical problem. Team members seemed to be reluctant in the early stages to get the technical-type problem solved in their home organization. This resulted in some procrastination and loss of time, but was not critical to achieving end results.

The fact that each group was placed together at one time produced conflict in the early stages in just plain getting along with one another. The group had to develop its own norms and rules and informal structure.

The uncertainties of the task itself tended to create some tension among team members. Although all experienced the same kind of tension, a greater number exhibited an ability to handle this than did not.

Some were concerned about rewards and accommodations in the earlier stages of the project. This again was a minority of members, but nevertheless the subject received abundant attention.

In addition to these, there was misunderstanding of matrix management, a concentration on the negative rather than positive aspects, belittling of the project manager, attempts to leave the project prematurely as noted earlier, concern about splitting the group in four separate geographic areas rather than one, and an attempt to meld the 18 members or a portion of them in one geographic location at a premature stage.

Driving Force. A specific log was not maintained concerning the frequent contact with individuals and teams that tended to either solve the above type problems or keep them at least neutral. All the action, however, can be classified as either confrontation with individuals when appropriate, working through key individual team members to solve team type problems, or confrontation with an entire team. The training consultant was utilized most effectively in all the areas noted above. He participated with the project manager in discussions with teams and had individual sessions with some team members that tended to keep the overall project objectives in accurate perspective.

Results. The four teams remained intact geographically throughout the entire year and accomplished all objectives. The specific results caused by the above action are too numerous and personal to mention. Most of the areas discussed, however, in each of the eight concerns had a bearing on the driving forces mentioned here.

Restraining Force. Low level of communication between teams was a serious problem from time to time throughout the project. This related to some of the restraining forces noted earlier, especially in Causes 4 and 5. This came about in the stages when plans were being made to gather data in the field, prepare the agenda for project meetings, exchange meaningful information concerning alternatives, make decisions concerning the level of testing required to meet project goals, and select a digest committee to prepare a summary report of project recommendations. All of these resulted in varying degrees of communication, but only reached maximum efficiency under pressure of one kind or another.

Another problem related to low level communication was the tentativeness of agreement between teams even when it was reached. Compatibility seemed to be lost at one stage and had to be bolstered through suggestions from the project manager for Field Teams A and C to become better coordinated.

Driving Force. The approach here was twofold. First, the communication problem had to be attacked by having the training consultants visit all field teams. The purpose of this kind of visit was to open up the problem of communication to determine why it existed and what could be done about it in general rather than any specific terms. The second manner of addressing the communication problem was to solve specific problems as they occurred.

Results.

- All project meetings with the exception of two were based on maximum communication between field teams, what subject should be discussed, the level of time involved, and whether the total group should be involved or subgroups with definition of specific subjects for both.
- As noted earlier, Field Teams A and C became compatible after they were asked to do so. This actually did occur in an efficient way. Another example is the communication to Field Team B by each of the other teams concerning their plan of attack on expanding final alternatives to more than one specific conclusion. This occurred.
- Each team tackled the testing problem in different ways using various approaches. A definition of the testing to be completed was confirmed with the sponsoring organization and communicated to each team. This aided the accomplishment of testing objectives both during the project and after.
- Based on some specific suggestions, each team changed its representative on the digest committee to a more appropriate person.

Restraining Force. The total group had an overall tendency to have meetings, make agreements, and then not follow through or take appropriate action in regard to those agreements. This relates to the last item discussed, namely, lack of or a low level of communication, but differs to the extent that this item relates to agreements being made at meetings and then not followed. This, in some cases, relates to poor communication, but cannot be considered as the chief primary reason.

Driving Force. Continuous follow-up was made to assure appropriate action at appropriate times concerning many subjects. The highlights of these are covered below under Results.

Results.

- The initial planning done during the training week to accomplish project objectives remained firm throughout the project. Detailed plans to arrive at this that were made by each team varied from following the general plan to ignoring it with varying degrees in between. The detailed plans, however, were originated, revised, and monitored so that overall project dates and objectives were, in fact, accomplished.
- The subject of what level of information was required at various stages got much attention and was agreed to at various times. In spite of

much discussion and many changes, the desired level was reached and a highly satisfactory report was given.

- As indicated earlier, the subject of testing got much attention and several attempts to define it failed. This required continuous effort and work with each team and the chain of command to arrive at a satisfactory and accepted definition.
- The final report format received much attention in the early stages of the project and was resolved for the most part when members from each field team met to define the actual format. This remained the same until the latter stages of the project when format only became important concerning the final digest report.
- Evidence of agreeing on a subject and not following through on the agreement was the consultants' role. This was explored in Cause 4 which indicated in detail the steps that were taken to alleviate the continued misunderstanding of the consultants' role. The overall result was effectuated in specific cases only, but remained a general problem throughout the life of the project.
- Another example of this phenomenon was the establishment of a digest group to work on recommendations of what a summary report should include. This went from an eight-person committee to a four-person committee; from a one-month project to a one-day project; from a relocation of people and families to no relocation. The basic agreement which was sound when originated and adhered to in spite of the above shifts was to come up with a summary report for the project including a description of alternatives and recommendations by the three teams. This was accomplished.

Restraining Force. Heavy travel, especially in the first part of the project, caused people to be away from home for extended periods on a repetitive basis. In some respects, this caused problems on the home front, the details of which are unknown.

Driving Force. A letter was originated from the general manager to the spouses of those involved in the project. This letter recognized the spouse's role as being an important one to the success of the project, and recognized the need for extensive travel on the part of every project member. This letter was well received.

Results. As indicated previously, the letters were well received and tended to ease frustration and aid in understanding the position of those involved in the project.

Restraining Force. There was a general feeling that "home" organizations were not considering performance ratings and promotional opportunities as they should. The distance from the "home boss" caused a feeling of being forgotten and tended to cause a gap concerning personal development discussions. This became known as the "out-of-sight out-of-mind syndrome."

Driving Force. The project manager had discussions with each functional boss prior to rating time and was assured that ratings either remained the same or were actually increased for all those involved in the project. This was communicated to individuals that seemed to have a high concern about rating treatment.

In addition, the project manager interviewed each individual concerning short-term goals, and, where appropriate, shared problems concerning growth with these individuals. Maximum use was made of the training consultant in this area, and interviews were set up with at least three of the project members. Where appropriate, discussions centered on submitting specific people for promotion. These cases involved persons of long service whose chances for promotion were slight even though they may have deserved it fully.

Results.

- All persons were treated fairly by "home" organizations regarding appraisal. No appraisal was changed downward. Some were changed upward.
- Communication of this to those individuals who were concerned seemed to help. The personal development discussions with the project manager and also the training consultant was a very positive action that led to better contributions by these individuals.

Cause 8

Level of Field Cooperation with Project Teams

This involves all organizations related to the project either directly or indirectly. This includes the full-time and part-time organizations. All organizations more or less responded to direct questions mainly through necessity of answering in a formal way; but two organizations very often initiated action on their own to involve themselves in a productive and positive way with the project.

In addition, the management training consultant involved himself in the project in the same manner as noted above; namely, responding when asked for advice and help, but also initiated positive action. The best example of this is the time spent with individuals in the project in helping them understand their current situation which helped them to contribute in a more positive way to the project. These sessions also helped individuals as far as short-term and long-term growth.

Restraining Force. The position of related organizations was, in most cases, to wait for the teams to contact them before responding. This was noted previously, and in summary relates to the problem of waiting rather than initiating action to help project teams when necessary.

Driving Force. Two-way communication was encouraged as being appropriate. This subject was stressed by the project manager and general manager to all part-time resource organizations. This aided in some ways, but for the most part it remained a project responsibility to get action from these organizations in one way or another.

When a particular organization reacted favorably, was a special help, or initiated action in one way or another, recognition was usually made by either the field team or the project manager, or even the general manager where appropriate. This tended to maintain that level of cooperation throughout the life of the project.

Results.

- Most field and headquarters organizations cooperated fully with the project teams in the many interviews that took place in the data-gathering stages of the project.
- The management Training Organization as noted was a consistent help to the project, to field teams, and to individuals — both in responding and initiating.

Restraining Force. Field Team B was the last of the project teams to get out in the field in the data-gathering stage. This was considered restraining for two reasons. One, the level of cooperation Team B would get from the field was unknown; and, two, their timing appeared to be off as far as initiating this kind of contact.

Driving Force. The project manager decided to hold off before applying pressure to Field Team B to make the field visits. The reasoning was that the team was spending a lot of time determining the correct approach and coming up with a questionnaire that would be all inclusive and would prevent the need for future follow-up visits. In other words, the team attempted to do a quality job in-house before making actual field visits.

Results.

- Field Team B completed plans for visitations and actually finished all visits in good time. These visits were well within project target dates, although it did put pressure on Team B in the development of the alternatives stage.
- Based on actual on-site visitations to field locations by the project manager with Field Team B and the favorable feedback to the general manager from key managers in the field, Team B exhibited maximum preparedness and, therefore, accomplished a great deal in each of its field visits. They enhanced the project image and accomplished the data-gathering phase in a most satisfactory manner.

Restraining Force. There was some field dissatisfaction with Field Team C's approach in the early stages of the data-gathering phase. The project manager received bad feedback from several sources that pointed out this dissatisfaction specifically.

Driving Force. This negative feedback was passed on to Team C by the project manager during a field visit. The project manager got the team's viewpoint on this and also checked, based on the team's suggestion, with an organization that had recently been interviewed.

Results. The overall feeling was that the Field Team C approach really contributed to the problem, and they revamped their plan of attack to eliminate any negative aspects in future visits.

Restraining Force. There was some need on the part of other organizations for more information concerning the project. This was mainly apparent in the early stages of the project.

Driving Force. It was decided that maximum saturation by verbal communication with all field organizations would be necessary. This was accomplished by the project teams, the project manager, and the general manager in conferences and also informal contacts, and also with the part-time resource organizations.

Results. Most organizations were cordial and cooperative in a meaningful way with the project teams.

Summary

Ninety-four results have been shared in this chapter. The idea that results can only be reported by use of numbers is a highly mistaken notion. John Naisbitt said the manager's new role is to coach, to teach, to nurture.[4] I would add "to facilitate." The results spelled out in this chapter are examples of what can happen when these traits are put to use.

Dr. Deming has said that the unknown and unknowable numbers are the ones that are most devastating. In this case it is not known what would have happened if action was not taken as illustrated in each driving force. It is highly probable that the project would have failed and the wealth of information, data, and recommendations would have been lost.

In this new economic age it has now become vital for managers to step back, take a hard look at what they are doing, and make rapid changes to ensure that pride returns to all places of work.

The recommendations of this project were accepted by the highest policy-making body over 15 years ago. The management process described was and still is avant-garde. We are now hearing phrases like partnershipping, servant leadership, and finding "moments of truth." These all refer to the kind of process we have been referring to throughout this book. Some will never understand or accept this process — others will. I hope the latter prevail over time.

Review

We have introduced a process that has helped improve quality in the AT&T Network Operations Organization in Atlanta and many other white-collar organizations throughout the company. This process, now being used by other companies, involves both workers and managers, and provides a road map of solutions to the problems that inhibit quality excellence. The end result of using CE/FFA lies in management *solutions* that have been well thought out and specify which levels of management need to take action (page v).

Part One — The Process.

- The CE/FFA technique allows those closest to the issues to identify problems, their causes, and the forces that can affect improvement (page 1).

Phase I.

- The purpose of this session is to help the management team understand its role in implementing and supporting quality improvement (page 1).
- The end result, however, is the same — trust your employees: believe in them; avoid overmanagement; drive out fear (page 1).
- We don't lose sight of the fact that higher management should get into the act on about 10 percent of the cases and employees can actually help in about 15 percent of the cases. The remaining 75 percent can be acted on by middle and lower levels of management. This is what I have found in 75 cases (page 1).
- To sum up, the purpose of *Phase I* is to get management's commitment to initiate QATs and QRTs and help them understand CE/FFA (page 2).

Phase II.

- In Phase II, I work with a small group (six to 10 people) and use the CE/FFA technique to explore a problem. The numbers are not important; Deming says it's 85/15; Juran says it's 80/20. In my work with 75 teams, it comes out 82/18. Remarkably close numbers (page 2)!
- It is wrong for managers to give employees the impression that they are responsible for only 15 percent of the problems. The employee may have a good idea about how to fix the process (which is in the 85 percent category) and management should encourage systems to get this input (page 3).
- We don't stop here though. It is very important to go to management with suggested solutions — not just problems. This is not a new theory. It's stated in every book on management ever written, but few people practice it (page 4).

- The session usually takes from six to 10 hours, based on the complexity of the problem. It accomplishes two objectives: (1) each member becomes trained on how to be a facilitator, and (2) a list of recommendations is formed for management action (page 4).
- Much has been written about the ideal size of a small group. Personally, I find less than six or more than 10 tends to be ineffective (page 5).
- Very often those closest to the work have very good ideas about what these solutions are and can make solid recommendations. Healthy disagreement sometimes leads to breakthrough in understanding a particular problem. Compromise may develop, which usually strengthens the situation (page 5).
- We usually recommend tackling the most difficult problems first. However, if one cause would take hours to solve and another months, get rid of the lesser one first (page 8).
- Use of CE/FFA demands a high level of attention and participation by the group. When levity can be used it is certainly encouraged. This relieves tension and stress and very often aids the group along the way (page 9).

Phase III.

- This is an exciting event. The workers get a chance to share their findings with the management team. (I have seen *six levels* together at the same time!) The beautiful part of this is the good quality presentation and the lack of extensive preparation or use of written reports. It's not teaching people how to talk — it's giving them a tool that allows them to express thoughts they know better than anyone else. I have seen hundreds of people do this with a zero percent failure rate. This covers a very wide spectrum of talent — from labor grades in installation to software developers with PhDs (page 10).

Phase IV.

- The purpose of this examination is to pinpoint areas of the process that can be measured, to build a data base in these measurements, to identify problem areas (potential or real), and to take action to correct and prevent these problems from occurring in the future (page 11).

Part Two — The Facilitator.

- All the reading and viewing videotapes cannot replace hands-on experience. We have found that the only way to train facilitators is to put them in a learn-by-doing environment. When they do it, they understand it (page 13)!
- The facilitator becomes the key in providing solid communication between workers and management concerning what needs to be done to improve the process (page 14).

- There is one thing that stands out as being most important. You must sincerely believe that the group you are working with knows more about the problem than you do. When you believe this, the rest falls into place. It's also best if you are not an expert on the problem being explored. As facilitator, your opinions are not appropriate in this forum. You are seeking the group's ideas (page 15).

Part Three — Use of Cause and Effect with Force Field Analysis.

- Effects are those quality problems or defects that occur on the specific job to which each group is assigned (page 17).
- The key is to put the right group of people together to examine that problem (page 18).
- After the group has determined the number one quality problem, the facilitator draws a cause and effect diagram on the pad and asks the group to suggest major causes of the problem to be studied (page 19).

Force Field Analysis.

- This concept states that any problem or situation is the result of forces acting upon it. Restraining forces are those that keep the problem situation at its current level. Driving forces are those that push the situation toward improvement. Restraining forces are the causes of problems; driving forces are the solutions (page 20).
- After the group has identified a significant number of restraining forces, it starts looking for driving forces to counter each specific restraining force (page 22).
- Once the group has identified all the restraining forces and their opposing driving forces for a specific major cause of a quality problem, it has completed a force field analysis (page 23).

Part Four — Anatomy of a Successful Field Trip.

- The talk went well. No talk is ever exactly the same. Many times I find myself reacting to a positive or negative gesture and inserting a comment that wasn't thought of previously. In this particular talk it appeared the group was seeking more definitions of Deming's "drive out fear" point (page 28).
- The other Deming point that many people challenge is the elimination of numerical goals and work standards. This group was no exception. Some say, "You only get what you measure." I would say, "And that's all you get!" Time and time again in my 30 years of experience I have seen people who could do more than they were asked to do (page 28).

- We had a situation where someone had the idea of motivating engineers to do a better job by drawing cartoons of engineers making mistakes. When management began to understand that the process the engineers were working on was the problem, not the engineers, the cartoons disappeared almost overnight. The important thing to understand from this experience is how to use the data when involving employees (page 29).
- Presentation to Management. I told the group that the presentation to management is an input session from the QAT to management. It's not the time for management to evaluate each statement or commit to specific action. This is the time to make sure management understands the input of the QAT and to clarify questions (page 41).
- I recommended that QRT be established with a supervisor as chairman. The third level manager volunteered, named his two immediate subordinates as members and asked for four volunteers from the QAT and one representative from the staff group (page 42).

Part Five — A Case Study

- I didn't know about cause and effect at that time. Now I can see that each of the problem areas defined were actually "causes" and the "effect" was that the overall project was in jeopardy. The cause and effect study is shown in Figure 18 (page 44).
- The point here is to explain the use of force field analysis, not to give definitions of the actual recommendations. Over time the recommendations may be augmented or changed completely; the use of force field analysis is the important thing for the reader to understand (page 44).
- As mentioned earlier, the training consultant suggested the force field analysis technique in analyzing various project manager concerns which helped determine the cause of certain problems and actions that could be taken to alleviate them. This contribution was considered vital to the overall success of the project (page 59).
- The aspect we are discussing here is the fact that the project members worked in an atmosphere where rank or structure of any kind was not recognized. It was, therefore, important that the impact of the project manager on the various teams be concentrated more toward the developmental mode of operation than either strict control or the opposite which would be relinquishing responsibility for reaching objectives (page 60).
- The definition of a synergistic approach is "the cooperative effort in which the effects are greater than the sum of each part (individual working alone)." This, in effect, is the major reason for lack of structure within the project itself, and securing people with the best talents to represent specific disciplines that could, in fact, meld together as

a team as well as a total group of 18 in adjusting themselves to the problem at hand and, in fact, solving it (page 60).

- The Theory Y approach mentioned earlier refers, of course, to McGregor's more enlightened view of the employee by management. This theory assumes that people generally seek responsibility; that they have a positive capacity for exercising imagination, ingenuity, and creativity; and that they can exercise self-control (page 60).

- A strict chain-of-command approach lessens exposure of the teams to needed input. The best example of this is access to higher levels of management for required input regarding past agreements, current activity, and interpretations of policy and procedures (page 62).

- There was a savings in time and money on the part of the project by making maximum use of even intuitive direction received as a result of all visits by higher management. The project manager in these cases utilized the technique of maximum communication between those who visited and those who were visited. This involved clarifying interpretations from both sides as to what was discussed, why, and what action should be taken (page 63).

- In all situations, there were attempts to accomplish results and relieve pressure on the teams at the same time. In many cases this clarified what was actually required and specific action or needs were communicated (page 64).

- *There was no dramatic difference, however, between the kinds of problems and confrontations that were experienced in reaching solutions during the first week of training and those encountered during the life of the project* (page 64).

- The major driving forces to counteract the above restraining type forces were to clarify the consultants' role continuously with the individual field teams when necessary, suggest action to the consultants individually or as a team when appropriate, support appropriate actions by either the field teams or the consultants concerning this specific problem, or attempt to provide constructive criticism when necessary (page 65).

- In addition to these, there was misunderstanding of matrix management, a concentration on the negative rather than positive aspects, belittling of the project manager, attempts to leave the project prematurely as noted earlier, concern about splitting the group in four separate geographic areas rather than one, and an attempt to meld the 18 members or a portion of them in one geographic location at a premature stage (page 66).

Closing Comments

To date I have worked with over 75 small groups and used the CE/FFA process successfully. I have used force field analysis to manage extremely complex projects as illustrated in Part Five. Although the process is useful in production-oriented situations I have found it to be extremely useful in the white-collar area. Accounting, billing, market planning, customer service, sales, legal, and public relations are examples.

Some questions were asked in the introduction that I would like to address specifically in these closing comments:

- *How do you get management's attention to act decisively in helping workers do a better job to break down barriers that prevent quality excellence?*

This is not a one-shot situation. It is good to gain common ground through use of the same language. This can be attained by attending the many courses that are available and by reading the books referred to in this book. It is everyday practice of the principles learned that is of utmost importance. Some have said, "Why does a company have to have its back against the wall before it gets into meaningful quality improvement." Unfortunately this is the case at times. But as Dr. Juran suggests it is easier to achieve actual improvement when your process is in control rather than trying to figure out what to do when the process is in chaos.

So how do you, as quality practitioner get management's attention? James Houghton, president of Corning Glass Works, suggests that you "discuss your cause, not your tools." He also advised," . . . Take the quality message to the top person in your organization as a *solution* to some of the organization's quality problems — take a position and point out the consequences of inaction."[15]

As cited earlier in this book, the top person need not be the CEO. Seek out what I call the location head — the person in charge of a specific function. You must have this person on your side. When I first started I called a friend who was location head. He cooperated fully and progress was attained. Then word began to spread to other organizations; people whom I have never met call me. The conversation is the same, however: "Please help me improve the quality of my operation." Then we go into the phases discussed in this book.

The use of employees and first-line supervisors on quality action teams followed by the use of quality resolution and review teams involving management definitely helps break down the barriers that prevent quality excellence. Workers end up doing a better job because the weaknesses in the process are identified and management does something to fix the process.

- *How do you get workers to participate in managerial decisions?*

John Akers, chief executive officer of IBM, discusses the importance of participative management in his talk at the American Electronics Association. As mentioned previously, John Naisbitt, Michael MacCoby, and Douglas

McGregor all discuss the importance of participative management for now and the future.

Use of the CE/FFA process described in this book gets worker participation in a disciplined way. It is better than MBWA, brainstorming, the Gordon technique, or cause and effect and force field analysis used separately. The combination is what gives the power to this process. Seventy-five groups have used this tool without failure. It has given workers and lower levels of supervision the great opportunity to have a say in what needs to be done to fix the process — and this is participation at its best. We merge these workers with management in our Phase III session where good participation usually takes place on the management side.

- *How do you get out there and ask the questions to get the workers' ideas crystallized?*

You use a trained facilitator in the CE/FFA technique. This is better than the casual conversation the boss gets involved in in walking around. Please don't misunderstand. Walking around may be fine but there must be more to get solid input. In the 75 sessions of six to 10 hours each, I have seen anywhere from 25 to 94 ideas created in each session. On the average this equates to over 2,500 recommendations — all aimed at fixing the process, not the air conditioning or cleaning windows or providing privileged parking or Sunday afternoon picnics or the texture of toilet paper! Some ask if I did a Pareto analysis on the 2,500 recommendations. The answer is no! These 75 groups all came from different organizations. The management of those organizations analyzed the recommendations to discover what they could do. I advise the management team that nothing in business can be considered *trivial*. If Ray Kroc believed this, McDonald's would have gone bankrupt 20 years ago. Clean trays, cooks wearing hats, and clean rest rooms are not trivial in the customer's eyes. Cleanliness and consistency of service and taste are the hallmarks of McDonald's worldwide success.

So, if we end up with a list of 30 recommendations, I recommend going after the easiest first. Why let something wait that can be fixed in 24 hours! Tackle the harder items later with a planned approach that will get the job done over a period of time. Assign someone responsible, agree on a target date, track that date, and record any possible cost savings.

- *How do you, as manager, unleash the worker's creative ideas to solve problems?*

First, give them time and opportunity to use the CE/FFA process with the help of a trained facilitator. Second, *listen* to the ideas, and third, take action. Unless this cycle is completed you will have done more harm than good. Dr. Deming has stated that American management doesn't want to step out and take a chance to help improve things. "It's easier to do nothing," he said.[16] We can no longer take this route. Management at all levels must now take ownership of the process and in one small step at a time act to improve it. I think we are closer to a now or never situation than ever before. Time is running out!

- *How does management set in motion a way to implement these ideas so that the work can be done better?*

This entire book is about showing "a way" to improve quality that works. If you have started your reading here, as I do occasionally, go back to page v and start at the beginning. You will discover "a way" that dramatically gets workers/supervisors and higher management together on preventing defects and controlling variability. The end result can be higher productivity, lower cost, and higher quality goods and services. This can result further in companies staying in business and providing more jobs. The latter, of course is Dr. Deming's formula. Dr. Deming made this statement in one of his lectures:

> You are living in the most underdeveloped country in the world — just think what the United States could do if every employee was allowed to make a maximum contribution and management managed the way they should.[17]

If this book just helps nudge us in the direction Dr. Deming has been pointing, it will have been worth the effort.

References

1. Blanchard, Kenneth and Spencer Johnson. *The One-Minute Manager.* New York: William Morrow and Co., Inc., 1982, p. 39.
2. MacCoby, Michael. *The Leader.* New York: Simon and Schuster, 1981, p. 52.
3. Peters, Thomas J. and Robert H. Waterman, Jr. *In Search of Excellence.* New York: Harper and Row, 1982, p. 1.
4. Naisbitt, John and Patricia Aburdene. *Re-Inventing the Corporation.* New York: Warner Books, 1985, p. 2.
5. Petersen, Donald. "Petersen: Operating Principles for Total Quality." *Quality Progress* 18, No. 4 (April 1985): pp. 42-43.
6. Olson, James e. "A Message from James E. Olson, Chairman, National Quality Month." *Quality Progress* 18, No. 10 (October 1985): p. 8.
7. Deming, W. Edwards. *Quality, Productivity, and Competitive Position.* Cambridge: Massachusettes Institute of Technology, 1982, p. 109.
8. Ishikawa, Kaoru. *Guide to Quality Control.* Tokyo: Asian Productivity Organization, 1983, pp. 18-26.
9. Lewin, Kurt. *Field Theory and Social Science.* Ed. D. Cartwright. Westport, Conn: Greenwood Press, 1975, p. 3.
10. McGregor, Douglas. *The Human Side of Enterprise.* New York: McGraw-Hill Book Co., Inc., 1960, pp. 33-34.
11. Juran, J. M. "The Quality Trilogy." *Quality Progress* 19, No. 8 (August 1986): pp. 19-24.
12. Scherkenbach, William W. *The Deming Route to Quality and Productivity.* Washington, DC: CEE Press, ASQC Quality Press, 1986, pp. 100-101.
13. Bradford, Leland P. *Making Meetings Work: A Guide for Leaders and Group Members.* LaJolla, Calif.: University Associate, 1976, pp. 42-50.
14. *AT&T Statistical Quality Control Handbook.* Indianapolis: AT&T Technical Publications, 1956, p. 9.
15. Houghton, James R. "A Message from the Chair of National Quality Month." *Quality Progress* 20, No. 10 (October 1987): p. 19.
16. *Road Map for Change/The Deming Approach.* Encyclopedia Brittanica Educational Corporation of Chicago, Part I, 1984.
17. The Deming Series of Lectures. Sponsored by Western Electric Company Merrimac Valley Works. Cape Cod, October 1983.

Index

ASQC 40th Annual Quality Congress, 13
AT&T Network Operations Organization, 75

Bell Telephone Laboratories, 2
Blanchard, Kenneth, v
Bradford, Leland P., 7

Cause and Effect, vi
  basics of, 5
  examples, v, 1-4, 17, 19, 23, 32, 44
Cause and Effect/Force Field Analysis (CE/FFA), vi, 1-3
  facilitator, in, 13
    support of, 13
    advice, 15
  fun, how to make, 9
  results, how to track, 9
  names, memorize, 5
  room setup, 5
  self-correcting, 23
  use of, 17
  videotape, 4
  white collar, in, 23
Common Causes, 2
Corning Glass Works, 81
Crosby, Phillip B., 1

Deming, W. Edwards, vi, 1
  85/15 rule 2, 3, 41
  proving 85/15 rule, 25
  formula, 2
  management support, lack of, 10
  NBC White Paper, 2
  numbers, unknown/unknowable, 73
  underdeveloped country, the most, 83
  The Deming Route to Quality and Productivity, (Scherkenbach), 3

Facilitator, 14
  support of, 14
  advice, 15
Feigenbaum, Armand V., 1
Field Theory and Social Science, (Lewin), 1
Force Field Analysis, vi
  basics, 6
  case study, 43
  examples, 20-22, 24, 33-40, 45, 47, 49, 52, 54, 56, 58-59, 69
Ford Motor Company, 28

Guide to Quality Control, (Ishikawa), 1, 3

Houghton, James R., 81

Ishikawa, Kaoru, vi, 1, 3

Juran, J. M., 1
Juran Impro Conference, 2, 28
  managers, limited experience and training, 1

Kroc, Ray, 82

Lewin, Kurt, vi, 1

MacCoby, Michael, v
Making Meetings Work: A Guide for Leaders and Group Members,
  (Bradford), 7
MBWA, v
Matrix Management, 59
McDonald's, 82
McGregor, Douglas, 1, 60
Melohn, Tom, 10

Naisbitt, John, v, 1, 73
NBC White Paper on Deming, 2
North American Tool and Die, 10

Olson, James E., v

Partnershipping, 73
Petersen, Donald, v
Peters, Thomas J., v

Quality Action Team (QAT), 1, 4, 11, 14, 42
Quality, Productivity, and Competitive Position, (Deming), vi
  QC circle, vi
  QC weakness of, vi
Quality Resolution Team (QRT), 1, 11, 42
Quality Teams — How Do They Work (Stratton), 4

Reinventing the Corporation, (Naisbitt), 1
Restraining Forces, 6
Rogers, Kip, 13

Scherkenbach, William W., 3, 28
Servant Leadership, 73
Shewhart, Walter A., 2
   control chart, his, 2
   rule of variability, his, 2
Special Causes, 3
Statistical Quality Control Handbook, (AT&T), 11
Synergistic Approach, 60

Theory Y, (McGregor), 60

Waterman, Robert M., v